CPS SERIES IN PHILOSOPHY OF SCIENCE

Center for Philosophy of Science
University of Pittsburgh

CPS PUBLICATIONS IN PHILOSOPHY OF SCIENCE

Center for Philosophy of Science
University of Pittsburgh

EDITED BY

Adolf Grunbaum
Larry Laudan
Nicholas Rescher
Wesley Salmon

CP

AESTHETIC FACTORS IN NATURAL SCIENCE

**Edited by
Nicholas Rescher**

UNIVERSITY PRESS OF AMERICA

Lanham • New York • London

University Press of America®, Inc.

4720 Boston Way
Lanham, MD 20706

3 Henrietta Street
London WC2E 8LU England

Printed in the United States of America

British Cataloging in Publication Information Available

CPS PUBLICATIONS IN PHILOSOPHY OF SCIENCE
Co-published by arrangement with the Center for Philosophy of Science
University of Pittsburgh

Library of Congress Cataloging-in-Publication Data

Aesthetic factors in natural science / edited by Nicholas Rescher.
p. cm.
(CPS publications in philosophy of science)
1. Science—Philosophy—Congresses. 2. Science—Methodology—Congresses.
3. Simplicity (Philosophy)—Congresses. I. Rescher, Nicholas.
Q175.A294 1989 501—dc20 89–35898 CIP

ISBN 0–8191–7576–5 (alk. paper)

The paper used in this publication meets the minimum requirements of American National Standard for Information Sciences—Permanence of Paper for Printed Library Materials, ANSI Z39.48–1984.

CONTENTS

PREFACE

These essays originated from an interdisciplinary conference on "Aesthetic Factors in Natural Science" held in Pittsburgh in December of 1987, under the sponsorship of the University of Pittsburgh's Center for Philosophy of Science. This conference, which was supported by the Richard King Mellon Foundation, brought together scholars from many disciplines in fruitful interaction. It is our hope that these publication papers will stimulate further reflections on this interesting topic which, of course, is so vast that the discussions presented here can do little more than scratch the surface.

AESTHETIC FACTORS IN NATURAL SCIENCE

Nicholas Rescher

I. Introduction

IN announcing the subject of the present conference, its agenda was stated in the following terms in our "Call for Papers": *The theme of the conference will be the examination and clarification of the place of "aesthetic" parameters in scientific explanation: simplicity, uniformity, symmetry, economy, elegance, and the like.* We had it in view that the range of deliberations would cover the whole spectrum of such aesthetic factors. At the back of our thinking was Rosalind Franklin's remark that the Crick-Watson double helix model "was just too pretty not to be right." However as fate and the spirit of the time would have it, just about every contributor chose to focus on simplicity.

On reflection, I came to feel that this is only natural. For at virtually every stage of the history of philosophy, those who have considered the issue have inclined to view these aesthetic aspects of scientific theorizing in much the same light, and to treat them in much the same way.

Let us make a brief Cook's tour of the historical situation. With Leibniz, the lawful order of nature—its economy, continuity, uniformity, simplicity, and the like included—represents a divinely ordained feature of this *actual* world, a feature which sets it apart from other *possible* worlds, reflecting the aesthetic concerns of the creator. For Kant, our commitment to the world's lawful order of aesthetic regularity reflects regulative principles of procedure rooted in the faculty structure of the human mind. For Hegel, the scientists' dedication to these structural factors of their world-picture is a social necessity ontologically rooted in the operations of the world spirit. For present-day empiricistically inclined theorists, our devotion to nature's simplicity, uniformity, and lawful orderliness simply reflects a social convention in the scientific community, whose basis lies in the pressure of habituation and custom—in the socially enforced mores of the community. Everyone lumps these factors together, though not, as I myself see it, in quite the right way.

The ensuing discussion will argue that a pragmatic approach in the spirit of Peirce offers us a viable middle way, between the sociological conventionalism of the moderns and the problematic metaphysics of their predecessors.

II. The Pragmatist View

Let me begin by sketching briefly—so briefly as to be almost a caricature—the sort of pragmatist approach to which I myself incline.

The approach agrees with Kant in viewing all the parameters of scientific systematicity—simplicity, uniformity, coherence and the rest—as methodological or procedural guidelines ("regulative principles"). And it agrees with the modern empiricists in viewing these guidelines as embodied in the conventional practices of the scientific community. But it refuses to see these conventional commitments as merely arbitrary customs—as resting on a purely sociological grounding. With Hegel, it insists on an underlying basis of rationality and thus rejects arbitrarist conventionalism. But it seeks to find its rationale-providing consideration not in the *global* rationality of the world spirit, but in the *local* rationality of efficiencies for the functional aims of objectives of the scientific enterprise. Our commitments to simplicity and its congeners are indeed customs, but they represent customary practices that bear the legitimating hallmark of a record of pragmatic efficiency.

That is, we treat simplicity, coherence, and the like as validating factors and grant precedence and preferability to theories that are comparatively better in these regards NOT because we have learned that simpler theories are thereby truer or more probably true, BUT RATHER because we have learned by *experience* that this practice is efficient (cost-effective) for the conduct of inquiry.

Let me set this position out a little more fully.

III. Induction and Cognitive Economy: The Economic Rationale of Simplicity-Preference

It has long been recognized that simplicity must play a prominent part in the methodology of science in representing a paramount factor in inductive reasoning. There is as widespread agreement as there ever is in such foundational matters on the principle that simple hypotheses enjoy a preferred status. But when one presses the question of *validating* this simplicity preference, one meets with discord and disagreement. After all, what good reason is there to think that nature inclines towards the simple?

The matter becomes far less problematic, however, once one approaches it from a methodological rather than a substantive point of view. The simplest feasible resolution of our problems must surely be allowed to prevail—at any rate *pro tempore*, until such time as its untenability becomes manifest and complications force themselves upon us. Where a simple solution will accommodate the data at hand there is, in the circumstances, no good reason for turning elsewhere. It is a fundamental principle of rational procedure, operative just as much in the cognitive domain as anywhere else, that among various alternatives that are anything like equally well qualified in other regards, we should adopt the one that is the simplest—the most economical in whatever modes of simplicity and economy are relevantly applicable. Nature may or may not favor simplicity, but *we* should certainly do so—simply as a matter of rational procedure.

In induction we exploit the information at hand to answer the questions in the most straightforward (economical) way. Suppose, for example, that we are asked to supply the next member of the series:

1,2,3,4, . . .

We shall straightaway respond with 5, supposing the series to be simply that of the integers. Of course, the "actual" series might well be:

1,2,3,4,11,12,13,14,101,102,103,104, . . .

And the "correct" answer will then eventuate as 11 rather than 5. Though we cannot rule such possibilities out, they do not deter our inductive proceedings. The "inductively appropriate" course lies with the production rule ("Add 1 to the number you've just produced") that is the simplest answer. In induction we proceed to answer questions by opting for the simplest resolution that meets the conditions of the problem. And we do this not because we know *a priori* that this simplest resolution will prove to be correct. (We "know" no such thing!) We adopt and accept this answer, provisionally at least, just exactly because this is the simplest, the most economical way of providing a resolution that does justice to the facts—and to the demands of the situation. We recognize that the possibilities exist but ignore them because there is no cogent reason for giving them favorable notice *at this stage*.

In inductive situations we are called on to answer questions whose resolution is beyond the reach of information at hand. To meet our cognitive objectives, we have to reach out beyond "the data." And we do this by projecting our problem-resolutions along the lines of least resistance. We try to economize our cognitive effort. We use the simplest workable means to our ends exactly because the others are more difficult to use. Whenever possible, we analogize the present case to other similar ones, because the introduction of new patterns complicates our cognitive repertoire. We use the simplest viable formulations because they are easier to remember and simpler to use. Insofar as possible we try to ease the burdens we pose for our memory (for information storage and retrieval) and for our intellect (for information processing and calculation). We favor uniformity, analogy, simplicity, and the like because that lightens the burden of cognitive effort. We avoid needless complications whenever possible because this is the course of an economy of effort. And just herein lies the justification of induction, for by its very nature induction affords us the most cost-effective—the economically optimal—means for accomplishing an essential cognitive task.

In the *Discorsi* Galileo wrote:

> When therefore I observe a stone initially at rest falling from a considerable

height and gradually acquiring new increases of speed, why should I not believe that such increments come about in the simplest, the most plausible way?[1]

Why not indeed? Subsequent findings may, of course, render this simplest position untenable. But this recognition only reinforces the stance that simplicity is not an inevitable hallmark of truth (*simplex sigillum veri*), but merely a methodological tool of inquiry—a guide post of procedure. When something simple accomplishes the cognitive tasks in hand as well as some more complex alternative, it is foolish to adopt the latter.

It is the universal practice in scientific theory construction when other things are anything like equal to give preference to

—one-dimensional over multi-dimensional modes of description
—quantitative over qualitative characterizations
—lower- over higher-order polynomials
—linear rather than non-linear differential equations

In each case, the former alternative is clearly "simpler" than the latter. To be sure—a lot of effort to the contrary notwithstanding—no theoretician and no philosopher has managed to provide an adequate *substantive* characterization of simplicity, answering to the formula

> X is simpler than Y iff they stand to one another in a relation of such-and-such a descriptive sort.

But a *methodological* or *procedural* characterization is something far easier to come by. The comparatively simpler is simply that which is easier to work with, which, overall, is the more economical to operate when it comes to application and interaction. Simplicity on such a perspective is a concept of the practical order, pivoting simply on being more economical to use—that is, less demanding of resources.

The ideas of economy and simplicity are the guiding principles of inductive reasoning. The procedure is that of the fundamental precept of rational procedure: "Resolve your cognitive problems in the simplest, most economical way compatible with an adequate use of the information at your disposal." Our penchant for simplicity is easy to justify on grounds of economy. If one claims that a phenomenon depends not just on certain distances and weights and sizes, but *also* (say) on such further factors as temperature and magnetic forces, then one must design a more complex data-gathering apparatus to take readings over this enlarged range of physical parameters. Or again, in a certain curve-fitting case compare the thesis that the resultant function is linear with the thesis that it is linear up to a point and sinusoidally wave-like thereafter. Writing a set of instructions for checking whether empirically determined point-coordinates fit the specified function is clearly a vastly less complex—and so more economical—process in the linear case.

In inductive reasoning we constantly make use of organizational principles for the structuring of our information: subsumptive classification schemes, connecting laws, coordinating analogies. All of these are means for the assimilation of given cases to general patterns that have an "aesthetic" appeal. All such instruments for developing information, processing it, and rendering it accessible are one and all means for the cost-effective resolution of our questions. Throughout, operating cost-effectiveness and inductive adequacy run hand in hand. Ease of operation—economy in brief—is the touchstone of our inductive praxis.

IV. The Methodological Aspect of Inductive Economy

On such a view, the issue of the aesthetic aspect of inductive systematization is best approached with reference not to *reality* as such, but to *our conception* of it—or rather, more accurately, to the ways and means we employ in conceptualizing it. Simplicity-preference (for example) is based on the strictly method-oriented practical consideration that the simple hypotheses are the most convenient and advantageous for us to put to use in the context of our purposes. There is thus no recourse to a substantive (or descriptively constitutive) postulate of the simplicity of nature; it suffices to have recourse to a regulative (or practical) precept of economy of means. For the parameters of inductive systematicity—simplicity, uniformity, regularity, normality, coherence, and the rest—all represent practical principles of cognitive economy.[2] They are labor-saving devices for the avoidance of complications in the course of our endeavors to realize the objects of inquiry. The rationale of simplicity-preference is straightforward. It lies in the single word *economy*. The simplest workable solution is through this very fact that which is the easiest, most straightforward, most inexpensive one to work with.

Simpler (more systematic) answers are more easily codified, taught, learned, used, investigated, and so on. In short, they are more economical to operate. In consequence, the regulative principles of convenience and economy in learning and inquiry suffice to provide a rational basis for systematicity-preference. Our penchant for simplicity, uniformity, and their "aesthetic" congeners is not a matter of a substantive theory regarding the nature of the world, but one of search strategy—of cognitive methodology. In sum, we opt for simplicity (and systematicity in general) in inquiry not because it is truth-indicative, but because it is teleologically cost-effective for the more efficient realization of the goals of inquiry.

From this perspective, then, simplicity preference emerges as a matter of *simplification of labor*, a matter of the "intellectual economy" of cognitive procedure. The penchant for inductive systematicity is a matter of striving for economy in the conduct of inquiry. It is governed by an analogue of Occam's razor—a principle of parsimony to the effect that needless complexity is to be avoided. Given that the inductive method, viewed in its practical and methodological aspect, aims at the most efficient and effective means of question-resolution, it is only natural that our inductive precepts should direct us always to begin with the most systematic, and thereby economical, device that can actually do the job at hand. Our systematizing procedures pivot on this injunction always to adopt the most economical (simple, general, straightforward, etc.) solution that meets the demands of the situation. The root principle of inductive systematization is the axiom of cognitive economy: *complicationes non multiplicandae sunt praeter necessitatem.* The other-things-equal preferability of simpler solutions over more complex ones is thus obvious enough: they are less cumbersome to store, easier to take hold of, and less difficult to work with. In sum, we seek simplicity not so much for its own sake—because of the aesthetics of the thing—but because this is cost-effective as a strategy for problem-solving.

V. Ontological Ramifications of Simplicity

But is there any reason to think that simpler theories have a better prospect of actually being true? Clearly there are difficulties here. Does nature exhibit a penchant for simplicity? Surely not: We cannot say, solely on the basis of general principles of some sort, that this world—the real world as such—must of necessity be a simple one. Nor is there any real need for doing so.

Hans Reichenbach has written:

> In cases of inductive simplicity it is not economy which determines our choice . . . (W)e make the assumption that the simplest theory furnishes the best predictions. This assumption cannot be justified by convenience: it has a truth character and demands a justification within the theory of probability and induction.[3]

This perspective seems to me gravely misleading. What sort of consideration would possibly justify the supposition that "the simplest theory gives the best predictions"? Any such belief is surely inappropriate. Induction with respect to the history of science itself—a constant series of errors of oversimplification—would soon undermine our confidence that nature operates in the way we would deem the simpler. The course of scientific history is a highly repetitive story of simple theories giving way to more complicated and sophisticated ones. The Greeks had four elements; in the 19th century, Mendeleev had some eighty; we nowadays have a vast series of stability states of variable durability. Aristotle's cosmos had only spheres, Ptolemy's added epicycles, ours has a virtually endless proliferation of complex orbits that only supercomputers can approximate. Greek science could be transmitted on a shelf of books; that of the Newtonian age required a room full; ours requires vast storage structures filled not only with books and journals, but with photographs, tapes, floppy disks, and so on. Of the quantities nowadays recognized as the *fundamental constants of physics*, only one was contemplated in Newton's physics, the Universal Gravitational Constant. A second was added in the nineteenth century, Avogadro's Constant. The remaining six are all creatures of twentieth century physics: The Speed of Light that represents the velocity of electromagnetic radiation in free space, the Elementary Charge, the Rest-mass of the Electron, the Rest-mass of the Proton, Planck's Constant, and Boltzmann's Constant.[4] It would be naive—and quite wrong—to think that the course of scientific progress is one of increasing simplicity. The pivotal role that science assigns to cognitive virtues like simplicity, coherence, systematicity, fecundity, symmetry, generality, and the like, on first view appears as problematic and question-begging. All too clearly, it is deeply problematic to stake a claim, tacit or otherwise, to any straightforward *ontological* linkage between simplicity, uniformity, and so on, and (probable) truth. It is only when we turn to the methodological standpoint of procedural economy that everything falls easily and naturally into place.

After all, we need not presuppose that the world somehow *is* systematic (simple, uniform, and the like) to validate our penchant for the systematicity of our cognitive commitments. Our striving for cognitive systematicity in its various forms—simplicity included—persists even in the face of complex phenomena: the commitment to simplicity in our account of the world remains a methodological desideratum regardless of how complex or untidy the world may turn out to be.

The pursuit of cognitive systematicity is thus ontologically neutral. It is noncommittal on matters of substance, merely reflecting the procedure of conducting our question-resolving endeavors with the greatest economy. In inquiry as elsewhere, a principle of least effort predominates—rationality enjoins us to employ the maximally economic means conducive to the attainment of chosen ends. Such an approach constitutes a *theoretical* defense of inductive systematicity that in fact rests on *practical* considerations.

It is important, however, to distinguish economy of means from economy of product—methodological from material economy. Simple tools or methods can, suitably used, create complicated results. A simple cognitive method, such as "trial and error" can ultimately yield complex answers to difficult questions. Conversely, simple results are sometimes brought about in complicated ways. A complicated method of inquiry or problem-solving might yield easy and uncomplicated problem-solutions. Our commitment to simplicity in scientific inquiry does not, in the end, prevent us from discovering whatever complexities are actually there.

To be sure, however, this procedural/methodological tale is not quite the *whole* story. There is also, in the final analysis, a substantive aspect to the matter of induction's justifications.

Our *intellectual* tastes (for simplicity, elegance, and so on, as we naturally construe these ideas) are, like our *physical* tastes (palatability), the product of evolutionary pressure to the prioritization of those things that work —that prove effective, and thus survival-conducive. The evolutionary aspect of our cognitive mechanisms assures the serviceability of the cognitive values we standardly invoke as effective conditions of adequacy for the substantiation of information. The presence of positive values in information enhances the utility of that information—not because nature is benign or because a preestablished harmony is at work, but because evolution—both biological and cultural—so operates as to assure an alignment here. Nature exacts its penalties for ineffectiveness, and evolutionary pressure to cost-effectiveness assures an inherent connection between functional adequacy and temporal survival. The evolutionary realities assure an important role for economic considerations in the theory of knowledge. Cost-effectiveness is inevitably coordinated with the implicit rationality of evolutionary process in virtue of the survival conduciveness of arrangements that represent efficient ways of using limited resources.

Moreover, the development of our cognitive methods through *rational* selection also plays a key role in this connection. The process of cognitive evolution so unfolds as to assure *the coordination of convenience with effectiveness*. For a process of *rational* selection is at work to support the retention, promulgation, and transmission of these cognitive resources that prove themselves effective in operation. The burden of this evolutionary argument is not biological survival. The point is not that in a niche-rivalry between Austere Simplifiers and Byzantine Complicators, the former will eventually *displace the latter biologically*, but rather that in a fair contest in a community of intelligent inquirers, the former will eventually *outdistance the latter epistemologically*, simply through the very fact that they perform in a more cost-efficient way. Not biological selection alone but cultural dominance among intelligent agents plays a crucial part in the development of our cognitive instrumentalities.

The justification of relying on simplicity in the pursuit of our cognitive affairs will *initially* rest on an essentially pragmatic basis. We are to prefer the optimally systematic (simple, uniform) alternative in the first instance because this is the most economical, the most *convenient* thing to do. But *ultimately* we do persist in this course because experience shows the utilization of such economical methods to be efficient—optimally cost-effective (relative to *available* alternatives) for the realization of the task. The regulative principles and procedures at issue in our "aesthetically" oriented inductive practices are ones whose legitimation lies in their being pragmatically retrovalidated through a demonstrated capacity to guide inquiry into successful channels.

The crux is that we ultimately learn *by experience* (and thus through inductive reasoning itself) how to accomplish our "aesthetically" guided inductive business more effectively. For induction is a self-improving process. Experience can itself teach us what sorts of ways of interpreting the fundamental procedural ideas of inductive practice (simplicity, conformity, generality, and the rest) can lead to improved performance in the transaction of our inductive business. By a cyclic feed-back process of variation and trial we learn how to transact our inductive business more effectively. *Economy and convenience* play the crucial pioneering role in initially justifying our practice of inductive systematization on procedural and methodological grounds. But, in their turn, the issues of *effectiveness and success* come to predominate at the subsequent stage of retrospective revalidation *ex post facto*. And the question of the seemingly "preestablished harmony" coordinating these two theoretically disparate factors of convenience and effectiveness is ultimately resolved on the basis of evolutionary considerations in the order of *rational* selection.[5]

Accordingly, while our commitment to the "aesthetic" parameters of inductive procedure should be viewed in the first instance as a matter of methodological convenience within the overall economy of rational inquiry, nevertheless, our reliance on them is not *totally* devoid of ontological commitments regarding the world's nature. For our wisdom-of-hindsight experience with induction also enters into its overall justification—and indeed crucially determines not *that*, but *how* we perform inductions. It is this fundamentally methodological perspective that seems to me best to rationalize our standard reliance on aesthetic parameters in our endeavors at scientific systematization.

University of Pittsburgh

NOTES

1. Galileo Galilei, *Dialogues concerning Two New Sciences*, tr. by H. Crew and A. de Salvo (Evanston, 1914), p. 154.

2. Kant was the first philosopher clearly to perceive and emphasize this crucial point:

> But such a principle [of systematicity] does not prescribe any law for objects . . .; it is merely a subjective law for the orderly management of the possessions of our understanding, that by the comparison of its concepts it may reduce them to the smallest possible number; it does not justify us in demanding from the objects such uniformity as will minister to the convenience and extension of our understanding; and we may not, therefore, ascribe to the [methodological or *regulative*] maxim ["Systematize knowledge!"] any objective ((or descriptively *constitutive*)) validity. (CPuR., A306 = B362.)

Compare also C. S. Peirce's contention that the systematicity of nature is a regulative matter of scientific attitude rather than a constitutive matter of scientific fact. Charles Sanders Peirce, *Collected Papers*, Vol. VII (Cambridge, MA; 1958), sect. 7.134.

3. Hans Reichenbach, *Experience and Prediction* (Chicago and London, 1938), p. 376. Compare:

> Imagine that a physicist . . . wants to draw a curve which passes through [points on a graph that represent] the data observed. It is well known that the physicist chooses the simplest curve; this is not to be regarded as a matter of convenience . . . [For different] curves correspond as to the measurements observed, but they differ as to future measurements, hence they signify

> different predictions based on the same observational material. The choice of the simplest curve, consequently, depends on an inductive assumption: we believe that the simplest curve gives the best predictions . . . . If in such cases the question of simplicity plays a certain role for our decision, it is because we make the assumptions that the simplest theory furnishes the best predictions. (*Ibid.*, pp. 375-76.)

4. See B. W. Petky, *The Fundamental Physical Constants and the Frontiers of Measurement* (Bristol and Boston, 1985).

5. Further considerations relevant to these issues are canvassed in the author's *Methodological Pragmatism* (Oxford, 1977) and *Cognitive Systematization* (Oxford, 1979).

THREE ARGUMENTS AGAINST SIMPLICITY

Kristin Shrader-Frechette

I. Introduction

MANY philosophers of science have realized that the rationality of preferring simple scientific theories could be explained in terms of pragmatic and economic constraints such as allocation of time and ease of testing.[1] It is not difficult to justify preferring simple theories on grounds of practicality.

A more difficult question is whether there is any ontological connection between simplicity and probable truth. Many scientists and philosophers argue or at least presuppose that simplicity is a criterion for assessing theories.[2] As Boyd puts it, simplicity is a criterion of theory acceptability.[3]

While simplicity obviously plays a pivotal role in science, I argue in this essay that one common type of simplicity, what I call "O-R simplicity," provides no epistemological basis for accepting or rejecting scientific theories, and that O-R simplicity has a much more limited role in generating hypotheses and theories. I argue that its use is more appropriate to the context of discovery than to the context of justification.

II. O-R Simplicity

To substantiate the claim that at least one type of simplicity ought to play no role in theory acceptance or rejection, but only in theory generation, we need to establish what we mean by simplicity. It has been defined in many ways, for example, in terms of informativeness, falsifiability, testability, high probability, and low probability.[4] All of these definitions have come under criticism, at least in part because it is not clear that simple theories are more informative than complex ones, or more falsifiable than complex ones, and so on.

The focus of my remarks, O-R simplicity, appears to be both less stipulative and more in keeping with many of our intuitions than a number of other types of simplicity. I call it "O-R simplicity" or "Occam's-Razor Simplicity" after the seventeenth-century formula of William of Occam. Following the discussion of Occam's Razor by J. J. C. Smart [1984], I maintain that, given two possible hypotheses or theories consistent with the facts, one hypothesis or theory has more O-R simplicity than another when it postulates fewer principles, laws, properties, or entities. O-R simplicity appears close to what Feinberg was getting at when he said that the quark model was unlikely to have captured the fundamental constituents of the universe because the different properties assigned to quarks, in order to account for observations of hadrons, were too numerous.[5] (Admittedly, there are a number of questions associated with whether one or another theory, in a given case, may be said to have more O-R simplicity; since I take it that the notion of Occam's Razor is somewhat understood, let us leave such questions till another time.[6])

To substantiate my claim that O-R simplicity can be used to generate hypotheses and theories, but ought not be used to accept or reject them, I use three arguments. (1) If scientists used O-R simplicity as a basis for accepting or rejecting hypotheses or theories, false or at least counterintuitive consequences would follow. (2) Cases alleged to involve theory acceptance on the basis of O-R simplicity are really cases of evaluation in terms of some other criterion, like explanatory power. (3) If scientists used O-R simplicity as a basis for accepting or rejecting theories, dangerous consequences could follow. I close by suggesting some of the ways in which scientists use O-R simplicity to generate hypotheses and theories.

III. The First Argument: False Consequences Would Follow

The first argument is that if scientists did use O-R simplicity as a basis for accepting or rejecting one of two theories equally consistent with the facts, then false or counterintuitive consequences could follow. The most worrisome of these consequences would be to render "single-factor" theories plausible wherever the empirical fit between theories and phenomena was poor. This could happen in areas of science where direct observation often is difficult, like high-energy physics, or in areas where successful prediction is almost impossible, like parts of psychology, or in areas of relatively new science, like much of ecology. When there are several empirically underdetermined theories, for example, regarding whether competition or predation structures communities in ecology, and when these are equally consistent with the facts, use of O-R simplicity could dictate choosing a crude, suspect, single-factor theory.

For example, it has been notoriously difficult to obtain a fit between economic phenomena and economic theory. This is in large part why Mill chose the crude single-factor assumption that economic behavior could be predicted on the basis of only one parameter: the human desire for wealth.[7] Now obviously all economic behavior cannot be predicted on the basis of this single factor, but it is not clear that any multi-parameter theory is a better predictor, especially for all cases. Hence on grounds of O-R simplicity, Mill's single-parameter theory appears justified, even though it is false.

From an epistemological point of view, there is no reason to suppose that, given two empirically underdetermined theories consistent with the facts, the theory with more O-R simplicity is more likely to be true. [See Maxwell, 1975, p. 159.] In fact, in sciences explaining human behavior, it might be more reasonable to suppose that theories possessing less O-R simplicity were more plausible, simply because they might account for more complex phenomena. If so, using O-R simplicity to assess theories in the social sciences could lead to counterintuitive consequences, for example, accepting explanations in terms of single factors.

More generally, as Friedman has pointed out, using simplicity to choose from among theories, all of which are consistent with the facts, is bound to lead to counterintuitive conclusions about the most acceptable scientific theory. This is because, for any such theory, there is a simpler one also consistent with the facts. Using Nelson Goodman's example [1961], we might say that the hypothesis, "all maples whatsoever, and all sassafras trees in Eagleville, are deciduous" is consistent with the facts. But another hypothesis, "all maples are deciduous," is also consistent with the facts, and it has more O-R simplicity, since it postulates fewer types of trees and makes no mention of properties of trees in Eagleville. Even though the second hypothesis would be chosen on grounds of O-R simplicity, obviously it is not the most acceptable scientifically since it is weaker and provides no account of sassafras trees. If Friedman is right, that science strives for strong, not safe, hypotheses, then using O-R simplicity to choose among theories could lead to accepting theories that are weaker or less acceptable than others, in terms of explanatory power.[8]

IV. The Second Argument: Scientists Rarely Use Simplicity

At this point, it might be objected that it is unfair to claim that on grounds of O-R simplicity, one would choose the hypothesis about maple trees over the hypothesis about maple and sassafras trees, since they do not have equal explanatory power. Rather, the objector might claim, O-R simplicity merely requires one to choose the simpler of two hypotheses, when both are equally able to explain the same facts. [See Schaffner, 1970, p. 326.]

First of all, I doubt that there are many situations in which two hypotheses are really *equally* able to account for the facts, even though several hypotheses may be consistent with the facts. If so, then the objection fails. It also fails because it presupposes that, when one evaluates hypotheses on grounds of simplicity, then one must take empirical criteria, like explanatory and predictive power, into account; once one claims that the O-R simplicity of several hypotheses ought never be evaluated, independent of empirical criteria like explanatory power, then it is questionable whether one can still claim to be evaluating *simplicity*. Instead, when scientists allegedly use simplicity in the context of justification, I suspect that they have really stipulatively defined it as something else, something related to empirical plausibility.[9]

For example, consider Schaffner's claim, that "one of the fundamental reasons why Einstein's theory attracted the attention and commitment of early twentieth-century physicists was because of its simplicity in comparison with Lorentz's theory."[10] Schaffner stresses the importance of Einstein's using only two postulates and eliminating the ether as examples of (what I have called) O-R simplicity. However, I'm not so sure that simplicity is what was at work here.

Einstein was not preferred primarily because he began with only a small number of postulates, the relativity principle and the light postulate, but primarily because these were two well supported postulates. Einstein was not preferred merely because he eliminated the ether, but because he had good empirical reasons for eliminating the ether: his reanalysis of the concept of simultaneity enabled him to show that there were no ontologically privileged times or length measurements associated with a special reference frame. In other words, simplicity was not the main grounds for preferring Einstein, but the fact that his simpler account was well supported. Were his simpler account not better supported than the Lorentz account, especially after then there would not have been grounds for preferring it. But if not, then it is questionable to say that it was preferred on grounds of simplicity. Likewise, the Lorentz account was not rejected mainly for not being simple, but on the grounds that the principles and properties postulated were not well supported.[11]

Scientists would be hard put to think of a case in which a complex theory was, or ought to have been, rejected on grounds of O-R simplicity even though the theory was well substantiated. If so, then this is a strong reason for believing, as Boyd [1985, p. 88] put it, that what we call "evaluation of theories in terms of simplicity" is really just a special case of evaluation in terms of evidential support. But if so, then O-R simplicity is, at best, a second-order criterion, to be used after first-order criteria such as explanatory power. It is true that false theories typically are more encumbered with *ad hoc* hypotheses, but this is not a problem with simplicity, but with explanatory power and testability. Moreover, simply because false theories often have little O-R simplicity does not mean that one ought to conclude that theories with more O-R simplicity are more likely to be true; the presence of O-R simplicity may or may not be associated with high explanatory power, testability, and so on.

V. The Third Argument: Simplicity and Dangerous Consequences

A third argument against use of O-R simplicity is that choosing/rejecting a theory on grounds of O-R simplicity, especially in areas of science that are highly empirically underdetermined, can lead to dangerous consequences. In such situations, it might be better simply to admit that there is no adequate theory; otherwise, use of O-R simplicity might give simple-minded theories an air of specious acceptability causing them to be misused, especially in cases involving public policy.[12]

Consider a little-known case, probably typical of the way that use of the criterion of O-R simplicity encourages both poor science and poor use of science as a basis for policy. Beginning approximately 25 years ago, there was a conflict between two groups of geologists evaluating groundwater-flow theories for a proposed radwaste site in Maxey Flats, Kentucky. The geologists from several universities (Georgia Tech and Auburn, among others) and from several consulting groups and industries (primarily EMCON and NECO) used an extremely simple, single-factor flow theory, premissed completely on the low permeability of the shale on site. They concluded that the radwaste could not migrate offsite for centuries. The geologists from several government agencies (USGS and EPA) rejected the simple flow theory of the academic and industry geologists, and claimed that many factors, such as possible fissures and fractures, and hairline cracks between bedding planes, not merely the low permeability of the shale, had to be taken into account. Offering a theory with less O-R simplicity, and more explanatory parameters and properties, they claimed that radwaste could well migrate offsite.[13]

In supporting the low-permeability theory, the academic and industry scientists alleged that exploratory drilling at depths of from 90 to 320 feet revealed no water that could act as a transporter of waste, and hence that "the possibility of subsurface migration off-site is nonexistent."[14] They also indicated that there was no perched groundwater in the geological formation (the Nancy-Member shale) used for the trenches, and that the Nancy Member was "essentially impermeable."[15] Since their laboratory tests substantiated their field tests, and laboratory tests indicated that hydraulic conductivity or permeability should permit trench water to move only one foot in 100 to 1000 years, EMCON-NECO scientists were confident in using a very simple, single-parameter hydrogeological theory for groundwater transport.[16]

In rejecting the theory that low permeability or hydraulic conductivity at Maxey Flats would prevent offsite migration, the USGS-EPA scientists' main argument was that the low-permeability theory could not explain certain anomalies, for example, well recharge and radiochemical evidence of subsurface movement between trenches and wells.[17] The low-permeability camp, however, rejected the USGS-EPA data, claiming that the anomalies were explicable in terms of evaporator pollution, not subsurface migration.[18]

Because groundwater flow on the site was so slow and so difficult to monitor directly, both theories were empirically underdetermined and hence equally consistent with the facts.[19] At least in part because both theories were consistent with the facts, policymakers chose the theory with more O-R simplicity, the low permeability theory of the industry and academic geologists. Their choice in terms of O-R simplicity was dangerously wrong, however, because only nine years after the facility was opened, plutonium was discovered two miles offsite and the facility became known as the world's worst nuclear dump. The moral of the story is probably that, the greater the empirical underdetermination of several theories, the more dangerous it is to evaluate them in terms of O-R simplicity. Such an evaluation could lead to acceptance, as in the geology case, of simple-minded "single-factor" theories both questionable in themselves and dangerous in situations of applied science.

But if use of O-R simplicity in accepting and rejecting theories can lead to false consequences or to dangerous applications of science, and if scientists are misled in thinking that they have evaluated theories in terms of simplicity when they have not, then does this mean that there is no role for O-R simplicity in science? It seems to me that scientists are misled, not about the importance of simplicity, but about the stage of science at which it ought to be employed.

VI. Scientists Use O-R Simplicity to Generate Hypotheses

There are grounds for believing that the criterion of O-R simplicity is quite useful in helping to generate hypotheses and theories, even though, for the reasons already given, it ought not be used for theory acceptance or rejection. Perhaps economy is the most important reason why scientists use O-R simplicity in formulating hypotheses. Physicist John Wheeler said much the same thing: that employing simple theories, postulating only a few entities, enables physicists to make and correct mistakes as rapidly as possible; they can typically test simple theories more quickly than complex ones.[20] If so, the scientist has something in common with the medical doctor and the auto mechanic.

In attempting to diagnose the problem with an ailing car, an auto mechanic typically proposes the simplest hypothesis first. That is, she typically checks the ailment that is easiest and cheapest to fix. Often what is easiest and cheapest to fix are problems involving only one malfunctioning component, rather than a multiplicity of them. If an automobile is overheating, for example, she might first check for water in the radiator or for a radiator leak. Only after she has checked simple, single-factor hypotheses would she typically move to multiple-parameter hypotheses, such as that the overheating was caused *both* by a faulty thermostat and a defective water pump. Purely for reasons of economy, medical doctors do the same thing; all things being equal, they hypothesize first about O-R simple, single-parameter explanations; only after those fail do they propose a more sophisticated, multiple-factor etiology for a particular ailment.

A similar situation occurs when scientists are forced to suggest new entities or properties, as in the case of Pauli's postulating the neutrino or quark theorists' postulating new properties to account for hadron "observations" at higher energies. They typically propose the least number of new entities or properties consistent with the observations or theories of which they must take account. Their simple hypotheses are not accepted or rejected, however, until there are good empirical reasons for doing so. This suggests that O-R simplicity gives us a good basis for making rapid progress toward scientific progress; it is a good vehicle for the logic of discovery.[21]

VII. Three Objections

In response to this account of the role of simplicity in scientific discovery and in the Maxey Flats geological case study, philosophers of science have made at least three objections. Let's examine them and responses to them.

The first objection, by Vaughn McKim and Michael Kelly, is that simplicity may not have played a role in the Maxey Flats decision; rather, the flawed policy appears to have been the result of politics and vested interests, rather than the consequence of conceptual considerations like simplicity.[22] There are three important responses to this objection. *First*, the hydrogeological studies cited earlier in section V of this essay showed that the USGS scientists specifically rejected O-R simplicity when they argued for the necessity of a multiple-parameter model for groundwater flow, and that the academic scientists specifically accepted O-R simplicity when they argued that their single-parameter model had not been contradicted by any empirical evidence. Hence, apart from whether politics also played a role in the Maxey Flats case, it is clear the O-R simplicity was a major justification for adopting the simpler model.

There is a *second* response to this first objection, however. Even if O-R simplicity had not played a role at Maxey Flats, it is arguably correct that an outside observer, looking objectively at the data from the two groups of scientists studying the site, could have employed O-R simplicity as a justification for siding with the decision by the academic and industry scientists, the decision later shown to be incorrect. This is because both theories were arguably equally consistent with the facts; hence one could claim that there were grounds for preferring the simpler theory. In other words, in order to prove that scientists ought not use O-R simplicity, especially in empirically underdetermined, applied situations, one need not show that doing so has led to disastrous consequences, whether in the Maxey Flats or any other case. Instead, one only need show that, *if* there were rational grounds for employing O-R simplicity, and *if* one did employ O-R simplicity, then doing so would lead to disastrous consequences. This I have shown.

Third, the objection also fails because, although politics obviously played a role in the Maxey Flats case, this fact does not refute my claim about the use of O-R simplicity. Rather, if the second response to this objection is correct, then use of O-R simplicity could have enabled a scientific criterion, O-R simplicity, to be the vehicle for politics, as Campbell Whitaker has noted.[23] A different criterion, for example, predictive power, arguably could not have been misused this way, since the theories did not differ in their short-term predictive power (less than ten years), and since the rejected USGS theory was obviously superior in its long-term predictive power (greater than ten years). But if other criteria could not have been misused as easily as O-R simplicity, as a vehicle for politics, then this suggests the very points I have argued. There is a problem with using O-R simplicity in situations involving justification or theory choice. Likewise there is a problem even with presupposing that O-R simplicity has some credibility in theory choice, since this presupposition could allow the criterion to assume some legitimacy in a given situation and hence to be misused as a *means* to political or vested *ends*/interests. If one wished to make a purely political decision, it would be far more defensible to allege that there were scientific grounds (i.e., O-R simplicity) for doing so than to allege that there were not. But if so, then any alleged scientific criterion susceptible to misuse might also be questionable on practical grounds. This is exactly my point in the third argument against O-R simplicity.

The second objection, by Adolf Grunbaum, is that theories or hypotheses, like Lorentz', may not have been rejected on grounds of empirical adequacy, rather than simplicity. This is because it is difficult to establish, says Grunbaum, that any theory or hypothesis has been rejected on the grounds of falsifiability, and that any *ad hoc* hypothesis has problems with explanatory power and testability, rather than simplicity. Popper has not provided grounds, says Grunbaum, for his claim that the Lorentz theory is more falsifiable with respect to at least one of his avowed standards of greater falsifiability.[24] Hence Grunbaum argues that the burden of proof is on the person who alleges that testability, rather than simplicity, is the more important basis for theory choice.

There are at least three important responses to Grunbaum's objection. *First*, it is not clear why the burden of proof should be on the person arguing for testability over simplicity, as Grunbaum claims, rather than on the person arguing for simplicity over testability. *Second*, in classic cases in the history of science, it appears that *ad hoc* hypotheses have been rejected on grounds of testability, rather than on the basis of simplicity. Cases that come to mind are the modifying ascription of negative weight to phlogiston, the initial postulation of the existence of neutrinos, and the initial postulation of the existence of virtual particles.[25] Admittedly, the postulation of these *ad hoc* hypotheses was often taken to deserve no censure, but merely the warning that independent empirical evidence was needed before the hypotheses could be taken seriously. These three cases suggest that testability is typically a sufficient condition for *confirmation*, although simplicity may be grounds for *postulation*, of a hypothesis or theory, as was argued in this essay. *Third*, Grunbaum himself admits that independent empirical support for an *ad hoc* hypothesis or competing theory is necessary before the theory ought to be accepted, despite the fact that he claims that Popper has not provided grounds for his claim that one theory is more falsifiable than another with respect to at least one of his avowed standards of greater falsifiability. Grunbaum writes: "Before *H* can be taken seriously as a cognitively viable alternative to some (tacit) contrary logical constituent of T_1, and thus before T_2 can be expected to remain a better theory [because it is simpler] than the refuted T_1, the procurement of "additional" or "*independent*" empirical support for *H* is desirable or needed."[26] But if so, then despite the simplicity of a given theory or hypothesis, Grunbaum admits that only testability or independent empirical support is grounds for accepting it or confirming it. And if so, then my claims about the role of O-R simplicity stand.

The third objection, by Nick Rescher, is that it is wrong to say that use of the criterion of simplicity caused the erroneous site decision at Maxey Flats. Rather, says Rescher, any criterion that recommended the site would be wrong; therefore this example does not count against simplicity any more than any other criterion.[27] On the contrary, however, as I argued in my third response to the first objection, the example does count against simplicity more than other criteria. This is because a different criterion, for example, predictive power, could not have been so easily misused so as to justify opening the site. As I argued earlier, this is because the theories did not differ in their short-term predictive power (less than ten years), and because the rejected USGS theory was obviously superior in its long-term predictive power (greater than ten years). But if predictive power could not have been misused as easily as O-R simplicity, then this suggests that Rescher might be wrong in alleging that the example counts no more against simplicity than against any other criterion, like predictive power. And if so, then there is likely a problem with using O-R simplicity, especially in applied situations involving justification or theory choice.[28]

University of South Florida

NOTES

1. Van Fraassen, 1980, p. 88; Rescher, 1976, esp. pp. 87-89; Boyd, 1985, p. 51.

2. See, for example, Bronowski [1951, pp. 130, 133], Feinberg [1977, p. 262], Frank [1957, p. 352], Hesse [1975, pp. 100-04], Jeffreys [1937], Kemeny [1953, 1955], Kordig [1971, Ch. IV], McMullin [1976], Quine [1966], Rolston [1976, pp. 438-440], Schaffner [1970; 1974], Schlesinger [1975], Sober [1975], Strawinski [1982], and Thagard [1978].

3. Boyd, 1985, p. 56.

4. See, for example, Sober, 1975; Popper, 1968; Friedman, 1972; Jeffreys, 1937; Fales, 1978; Kemeny, 1955; and Schaffner, 1970, 1974.

5. 1977, p. 262. On Feinberg's view, a quark theory postulating two properties of quarks would probably would be said to have more O-R simplicity than a quark theory postulating

four properties, all other things being equal (which they never are). What I call O-R simplicity seems close to what Schaffner calls "ontological simplicity." He distinguishes between system simplicity and ontological simplicity [1970, p. 328]. See also Maxwell, 1979.

6. Admittedly, there are likely to be disputes over whether one theory is superior to another in terms of O-R simplicity, especially if, for example, one theory postulates fewer properties or entities than another, but these few properties are more complex than those of other theories. Because of time constraints, let us leave aside such questions of how to assess O-R simplicity and assume that, at least in simple cases, we understand it. Just as we can determine which theories are metaphysically sparse and which are luxuriant, since we understand Occam's Razor, so also there are many cases in which we can determine when one theory possesses more O-R simplicity than another. For a possible test of O-R simplicity, see J. Cornman, 1980, pp. 225-51. See also Thagard, 1978, pp. 88-89.

Part of the difficulty associated with determining whether one theory has more O-R simplicity than another is that the historical O-R principle has been subject to considerable controversy. It is possible, for example, that Occam did not subscribe to O-R simplicity or to parsimony as a metaphysical principle concerning numbers of entities, but that he did hold it as a methodological principle, although he was not the first to do so. See, for example, Roger Ariew, 1977.

7. Mill, 1985, pp. 52-54.

8. Friedman, 1972, p. 25.

9. See Boyd, 1985, p. 89.

10. Schaffner, 1974, p. 73.

11. Schaffner, 1974, p. 73. Zahar, 1973, argues that Einstein's program superseded Lorentz' in an empirical sense in 1915 with its explanation of the precession of Mercury's perihelion. More generally, Zahar maintains that Einstein's program had greater heuristic power than Lorentz'. Schaffner maintains, however, that many scientists became convinced of the superiority of Einstein's program because of "trans-empirical considerations," like simplicity. For Zahar, simplicity is derivative to empirical considerations. Note that in an earlier essay, Schaffner [1970, p. 330] argued that theoretical context considerations outweighed simplicity. For a discussion of the Einstein-Lorentz conflict, see Grunbaum, 1976, esp. pp. 336, 340, who argues that the Lorentz theory was rejected because it "would fail to secure subsequent independent experimental confirmation as against the claims of a new rival theory." See also note 24 below.

To some extent, the question of the role of simplicity in the choice between Einstein and Lorentz is a matter of the time about which one is speaking. As Hesse [1975, p. 103] notes, in the period immediately following Einstein's 1905 paper, Lorentz' theory was consistent with all known empirical results. As time went on, however, Einstein's theory gained empirical superiority. See Hesse 1975, p. 103, for a good bibliography on the simplicity comparisons of Einstein's and Lorentz theories.

12. Single-factor theories in psychology could be notoriously misleading and harmful if they were used to proscribe certain sorts of behavior, for example, in the way that Freudian theories were used earlier in this century to proscribe the behavior of suffragettes. Other single-factor psychological theories, such as those used in risk assessment, are equally dangerous. According to one school of risk assessors, risk aversion is linearly related to average annual probability of fatality. This simple, single-factor, linear hypothesis has

been used by many members of the risk-assessment community to condemn much of the risk behavior of the public. Starr, Whipple, Okrent, Maxey, Cohen, Lee, and others have alleged that if a person's risk aversion is not directly proportional to probability of fatality from that risk, then the person is irrational and inconsistent.

While it is obvious that factors like equity of risk distribution and benefits to be obtained from taking the risk, and not merely probability of fatality, have a bearing on risk aversion, there is no specific hypothesis or law more consistent with the facts than the simple linear hypothesis. Hence, even though it is obviously incorrect, it is preferred on grounds of O-R simplicity.

13. After EMCON and NECO scientists presented their simple low-permeability models and argued for the impossibility of off-site trench-water transport, the USGS Maxey Flats project director explicitly said that any simple quantitative model for the site was impossible because Maxey Flats is a poorly permeable, fractured geological system. Any prediction of flow paths would be impossible, he said, because of the highly irregular hairline fractures and fracture intensities [Zehner, 1981, pp. 35, 40; see Werner, 1980, p. 45.] He also argued that the inhomogeneity of the flow system was further illustrated by the differences in the radiochemical quality of the drilling and well samples. [Zehner, 1981, p. 153.]

Because one would need detailed information about the spatial distribution and hydraulic properties of each of many hairline fractures in a variety of successive strata (e.g., Nancy Shale, Farmers Sandstone, Henley Shale, Sudbury Shale, Bedford Shale, Ohio Shale, and Crab-Orchard Shale), data that is "rarely available," the USGS project director concluded that groundwater flow at Maxey Flats could not be predicted. [Polluck and Zehner, 1981, p. 3.] Any model, he said, would presuppose conditions for its accuracy that were not met at Maxey Flats, conditions such as "uniform transmission of water through the rocks." [Zehner, 1981, p. 132.] In sum, he claimed that "hydraulic conditions do not meet the requirements for the method of analysis." [Zehner, 1981, p. 134.] No model could be predictively accurate and, even if it were, there would be no way to check the predictions.

Despite the implausibility of any model for the site, the Project Director attempted to give an approximate upper limit of 50 feet per year for groundwater velocity. [Zehner, 1981, pp. 161-62.] He therefore implied that one ought to remember that the predictive accuracy of any groundwater model for the site was suspect and hence that one ought not weigh the criterion of simplicity too heavily in attempting to superimpose some simple model on the site. [Zehner, 1981, p. 19.] Even to arrive at his own model for possible maximum groundwater movement, he admitted that he had to make many assumptions not capable of being checked predictively, e.g., that the horizontal conductivity was 100 times larger than the vertical conductivity at the site [Polluck and Zehner, 1981, p. 7.]

14. Neel, 1976, p. 258.

15. Zehner, 1981, p. 27; Hajek, 1976, pp. 272-73; EMCON, 1975, p. IV-9.

16. Neel, 1976, p. 258. Evaluation of permeability theories in terms of O-R simplicity was reasonable since it was difficult to obtain empirical information that the simple theories were incorrect. As one geologist put it: "It would be extremely difficult, if not impossible, to monitor contamination of ground water at Maxey Flats by wells located outside of the burial pits" because the rate of discharge is so low and because the path of flow is unknown. [Walker, 1962, p. 2.]

Another reason for the EMCON-NECO scientists' confidence in their simple hydrogeological models was that they adhered to the then-current theory that plutonium could not migrate through subsurface routes. Hence they were able to discount USGS-EPA claims to have found offsite plutonium arising from subsurface migration; they concluded, instead, that "the off-site monitoring has shown no detectable migration of radionuclides from the burial ground"; they reasoned that the plutonium must have come from surface migration caused by transport spills on site. [Kentucky Science and Technology Commission, 1972, p. 9.] In other words, the EMCON-NECO-Kentucky scientists appeared confident in heavily weighting the value of O-R simplicity, since the laboratory and field tests had shown that the shale was "essentially impermeable," and hence capable of being modeled in a very simple way. Even the Environmental Protection Agency admitted that the trench shales were "very impermeable." [EPA, 1973, p. 133.] Hence although they lacked "quantitative information" which could show the predictive accuracy of their theory, they believed that their simple theory was as consistent with the facts as the more complex theory that alleged that prediction of groundwater flow was impossible.

17. Weiss and Columbo, 1980, p. 77; Zehner, 1981, p. 106.

18. The EMCON-NECO scientists rejected the USGS and EPA data, alleging that it was inconsistent with the fact that mean concentrations of radionuclides in streams and stream sediment increased or decreased according to whether the evaporator (for contaminated trench water) was being used or not being used. [Clark, Interview 1986, 1986; Zehner, 1981, p. 56.] Other evidence that the low-permeability theories could be correct was the fact that mean concentrations of radionuclides in streams and stream sediment decreased after site management took precautions to insure that contaminated runoff did not pollute nearby streams. [Cohen, Letter to Rowe, February 26, 1976, p. 268.] Even the USGS scientists themselves admitted that the major source of the tritium found in the streams was from the trench-water evaporator, and that most of the other radionuclides apparently came from surface runoff, not from bedrock flow. [Zehner, 1981, pp. 58, 61, 108.]

19. EMCO-NECO scientists had strong empirical grounds for believing that permeability was low and therefore for believing that the groundwater system was well understood and for relying heavily on the criterion of *simplicity* in theorizing about possible groundwater movement. Yet, because they were relying on quite different field results, the USGS-EPA scientists charged that the EMCON-NECO theory was unable to explain several anomalies such as baseflow at Rock Lick Creek and well recharge; hence USGS-EPA scientists decided to reject the simple account of the EMCON-NECO scientists. The important point, however, is that both sides had data consistent with their findings and inconsistent with the findings of their opponents. Hence it would be quite reasonable to claim that both the low permeability theory and the high permeability theory were consistent with the facts.

20. Gordon Fleming, physicist at Pennsylvania State University, attributed this remark to John Wheeler in a remark at the University of Pittsburgh on December 11, 1987.

21. There is some evidence that Maxwell, although he believed that simplicity gives us clues to the truth, recognized that simplicity was more appropriate to the logic of discovery than to the logic of acceptance. He writes [1979, p. 642]: "All this shows the immense heuristic power of the simplicity thesis . . . I conjecture that theoretical physicists in practice make constant use of this heuristically powerful, indispensable thesis in their search for new theories."

22. Vaughn McKim made this objection on October 13, 1987, when a similar paper was

presented at the University of Notre Dame. Michael Kelly also made the objection, on February 10, 1988, when an earlier version of this paper was presented at the University of South Florida.

23. Campbell Whitaker made this point on February 10, 1988, when an earlier version of this paper was presented at the University of South Florida.

24. Adolf Grunbaum made this point in a conversation with me on December 11, 1987, at the University of Pittsburgh. For substantiation of his point, see Grunbaum, 1976. Grunbaum argues in this essay that, contrary to Popper's original claim, the Lorentz-Fitzgerald theory was not *ad hoc*, except perhaps initially [Grunbaum, 1976, p. 341]; hence he claims that philosophers ought not maintain that the Lorentz theory ought to have been rejected on the grounds that it was not testable in any sense. Grunbaum claims that the contraction hypothesis does have falsifiable consequences [Grunbaum, 1976, p. 343]. Moreover, contrary to Popper, Grunbaum argues that the Lorentz theory does not provide an example of Grunbaum's degrees of *ad hocness* when these are construed as related inversely to degrees of falsifiability (p. 343). This is because the Lorentz theory is not more falsifiable than the original ether theory. And it is not more falsifiable than the original ether theory because the two theories are incomparable with respect to each of Popper's avowed standards of rank-ordering theories according to greater or less falsifiability. Such content incomparability obtains, says Grunbaum, because the two theories are logically incompatible and yield contrary empirical predictions for the outcome of the Michelson-Morley experiment [Grunbaum, 1976, p. 345]. Thus, he concludes that there are no increases of content that could possibly furnish a sufficient condition for the Lorentz theory to be more falsifiable than the original ether theory. Hence Grunbaum concludes that, contrary to Popper's allegation, his version of a falsificationist criterion for the admissibility of auxiliary hypotheses and theory replacements does not vindicate the transition from the ether theory to the Lorentz theory. This is because Popper has not provided grounds, says Grunbaum, for his claim that the Lorentz theory is more falsifiable with respect to at least one of his avowed standards of greater falsifiability [Grunbaum, 1976, p. 346].

25. For discussion of the problems with virtual particles, see Shrader-Frechette, 1977, pp. 419-21.

26. Grunbaum, 1976, p. 330.

27. Nicholas Rescher made this point in conversation with me at the University of Pittsburgh on December 11, 1987.

28. The author is grateful to the National Science Foundation for an award which enabled her to accomplish this and other work on philosophical problems in hydrogeology. She also thanks Gordon Fleming, Adolf Grunbaum, Michael Kelly, Nicholas Rescher, and Campbell Whitaker for constructive criticisms of an earlier version of this essay. Whatever errors remain are the sole responsibility of the author.

REFERENCES

Roger Ariew, "Did Occam Use His Razor?" *Franciscan Studies*, vol. 37 (1977), pp. 05-17.

R. N. Boyd, "Observations, Explanatory Power, and Simplicity," in P. Achinstein and O. Hannaway (eds.), *Observation, Experiment, and Hypothesis in Modern Physical*

Science (Cambridge: MIT Press, 1985), pp. 47-94.

Jacob Bronowski, *The Common Sense of Science* (Cambridge: Harvard University Press, 1953).

D. Clark, "Interview, August 11, 1986, with K. S. Shrader-Frechette, in his Frankfort government office," unpublished notes, p. 4.

B. Cohen, "Letter to W. Rowe, February 26, 1976," in U.S. Congress, 1976, pp. 267-68.

James W. Cornman, *Skepticism, Justification, and Explanation* (Dordrecht: Reidel, 1980).

EMCON Associates and Jack McCollough, *Geotechnical Investigation and Waste Management Studies, Nuclear Waste Disposal Site, Fleming County, Kentucky, Project 108-5.2*, unpublished report, available from Emcon, 326 Commercial Street, San Jose, California, February 6, 1975.

U.S. Environmental Protection Agency, *Report to Congress on Hazardous Waste Disposal* (Washington, D.C.: Government Printing Office, June 30, 1973).

Evan Fales, "Theoretical Simplicity and Defeasibility," *Philosophy of Science*, vol. 45 (1978), pp. 273-88.

Gerald Feinberg, *What is the World Made Of?* (Garden City, New York: Doubleday, 1977).

Phillipp Frank, *Philosophy of Science* (Englewood Cliffs, NJ: Prentice-Hall, 1957).

K. Friedman, "Empirical Simplicity as Testability," *British Journal for the Philosophy of Science*, vol. 23 (1972), pp. 25-33.

Nelson Goodman, "Safety, Strength, Simplicity," *Philosophy of Science*, vol. 28 (1961), pp. 150-51.

Adolf Grunbaum, "*Ad Hoc* Auxiliary Hypotheses and Falsificationism," *British Journal for the Philosophy of Science*, vol. 27 (1976), pp. 329-62.

B. F. Hajek, "Letter to H. G. Holton of NECO, February 9, 1976," in U.S. Congress, 1976, pp. 272-74.

Mary Hesse, "Bayesian Methods and the Initial Probabilities of Theories," in Grover Maxwell and Robert M. Anderson (eds.), *Induction, Probability, and Confirmation*, Volume VI, Minnesota Studies in the Philosophy of Science (Minneapolis: University of Minnesota Press, 1975), pp. 50-105.

Harold Jeffreys, *Scientific Inference*, 3rd ed. (Cambridge: Cambridge University Press, 1973).

J. G. Kemeny, "Two Measures of Complexity," *Journal of Philosophy*, vol. 52 (1955), pp. 722-33.

----------, "The Use of Simplicity in Induction," *Philosophical Review*, vol. 62 (1953), pp. 391-408.

Kentucky Science and Technology Commission, "Technical Review of the Maxey Flats Radioactive Waste Burial Site," unpublished report, Frankfort, Kentucky: Kentucky Department of Human Resources, July 1972.

Carl Kordig, *The Justification of Scientific Change* (Dordrecht: Reidel, 1971).

Grover Maxwell, "Induction, Simplicity, and Scientific Progress," *Scientia*, vol. 114 (1979), pp. 629-53.

----------, "Induction and Empiricism," in Grover Maxwell and Robert M. Anderson (eds.), *Induction, Probability, and Confirmation*, Volume VI, Minnesota Studies in the Philosophy of Science (Minneapolis: University of Minnesota Press, 1975), pp. 106-

65.

Ernan McMullin, "The Fertility of Theory and the Unit for Appraisal in Science," in R. Cohen *et al.* (eds.), *Essays in Memory of Imre Lakatos* (Dordrecht: Reidel, pp. 395-432).

J. S. Mill, "On the Definition and Method of Political Economy," in D. Hausman (ed.), *Philosophy of Economics* (Cambridge: Cambridge University Press, 1984), pp. 52-69.

J. Neel in U.S. Congress, 1976, pp. 255-65.

D. W. Polluck and H. H. Zehner, "A Conceptual Analysis of the Ground-Water Flow System at the Maxey Flats Radioactive Waste Burial Site, Fleming County, Kentucky," USGS Open-File Report, 1981. Published in C. Little and L. Stratton (eds.), *Modeling and Low-Level Waste Management*, ORO-821, National Technical Information Service, U.S. Department of Commerce, Springfield, Virginia, 22161, pp. 197-213.

Karl Popper, *The Logic of Scientific Discovery* (London: Hutchison, 1968).

W.V.O. Quine, "Simple Theories of a Complex World," in Quine, *The Ways of Paradox* (New York: Random House, 1966), pp. 242-46.

Nicholas Rescher, "Peirce and the Economy of Research," *Philosophy of Science*, vol. 43 (1976), pp. 71-98.

Holmes Rolston, "A Note on Simplicity . . . ," *Philosophy of Science*, vol. 43 (1976), pp. 438-40.

K. F. Schaffner, "Einstein Versus Lorentz: Research Programmes and the Logic of Comparative Theory Evaluation," *British Journal for the Philosophy of Science*, vol. 25 (1974), pp. 45-78.

----------, "Outlines of a Logic of Comparative Theory Evaluation . . . ," in Roger H. Stuewer (ed.), *Historical and Philosophical Perspectives of Science*, Volume V, Minnesota Studies in the Philosophy of Science (Minneapolis: University of Minnesota Press, 1970), pp. 311-53.

George Schlesinger, "Confirmation and Parsimony," in Grover Maxwell and Robert M. Anderson (eds.), *Induction, Probability, and Confirmation*, Volume VI, Minnesota Studies in the Philosophy of Science (Minneapolis: University of Minnesota Press, 1975), pp. 324-42.

Kristin Shrader-Frechette, "Atomism in Crisis: An Analysis of the Current High Energy Paradigm," *Philosophy of Science*, vol. 44 (1977), pp. 409-40.

J.J.C. Smart, "Occam's Razor," in J. H. Fetzer (ed.), *Principles of Philosophical Reasoning* (Totowa, New Jersey: Roman and Allenheld, 1984).

Elliott Sober, *Simplicity* (Oxford: Clarendon Press, 1975).

W. Strawinski, "A Formal Definition of the Concept of Simplicity," in W. Krajewski (ed.), *Polish Essays in the Philosophy of the Natural Sciences*, Dordrecht: Reidel, 1982, pp. 187-96.

P. Thagard, "The Best Explanation: Criteria for Theory Choice," *The Journal of Philosophy* LXXV No. 2 (February 1978): pp. 76-92.

U. S. Congress, *Low-Level Radioactive Waste Disposal*, Hearings before a subcommittee of the Committee on Government Operations, House of Representatives, 94th Congress, Second Session, February 23, March 12, and April 6, 1976, Washington, D. C., U. S. Government Printing Office, 1976.

B. Van Fraassen, *The Scientific Image*, Oxford: Clarendon Press, 1980.

I. Walker, *Geological and Hydrologic Evaluation of a Proposed Site for Burial of Solid Radioactive Wastes Northwest of Morehead, Fleming County, Kentucky*, U. S. Geological Survey, Kearney, New Jersey, September 12, 1962, unpublished report. On file in the Louisville, Kentucky office of the USGS.

A. Weiss and P. Columbo, *Evaluation of Isotope Migration—Land Burial*, NUREG/CR-1289 BNL-NUREG-51143, U. S. Nuclear Regulatory Commission, Washington, D. C., 1980.

E. Werner, *Joint Intensity Survey in the Morehead, Kentucky Area*, prepared for the USGS by Environmental Exploration (consultants), Morgantown, West Virginai, unpublished study, Louisville, Kentucky, USGS, Water Resources Division, 1980.

Eli Zahar, "Why Did Einstein's Programme Supersede Lorentz'?" *The British Journal for the Philosophy of Science* 37 (1973): pp. 95-123 and 223-62.

H. H. Zehner, *Hydrogeologic Investigation of the Maxey Flats Radioactive Waste Burial Site, Fleming County, Kentucky*, Open-File Report, USGS, Louisville, Kentucky, draft, 1981.

SIMPLICITY AND THE AESTHETICS OF EXPLANATION

Joseph C. Pitt

UNLIKE paintings and other artistic expressions, when it comes to scientific theories it is difficult to say that beauty is merely in the eye of th beholder. For despite the fact that we disagree sometimes about the relative aesthetic merits of one theory as compared to another, nevertheless, we do have a sense of when a theory is simple or elegant or economical. This ability to recognize aesthetic features of theories reflects a consensus concerning the goal of scientific theorizing, which is explanation. On the other hand, when it comes to paintings it is not clear what the goal is; there could be several, and with respect to some of them, such as expressing the emotions of the painter, it may not be possible in principle to construct criteria by which to determine if the painting meets that objective. It is this lack of specificity of goals and criteria for determining degrees of success that underlies the inherent subjectivity of aesthetic judgment in the arts and appears to differentiate the aesthetics of art from the aesthetics of science.

Since the goal of scientific theorizing is the development of good explanations, it is important to separate criteria of adequacy of a good explanation from questions of the aesthetics of explanation. For it is not at all clear that an aesthetically pleasing explanation is a good explanation. While explanations may be assessed aesthetically, such considerations do not have an obvious bearing on the quality of an explanation *qua* explanation.

We can specify quite clearly what it means to have a theory which provides good explanations by appeal to such pragmatic considerations as prediction and manipulative control of the environment. However, and this is where the trouble starts, we also we seem to have a built-in bias which assumes that explanations with certain aesthetic properties have a greater likelihood of achieving that kind of pragmatic success. Thus a simpler explanation is said to be better than a more complicated one because it is dubiously assumed to have a greater likelihood of achieving the practical consequences we seek. This dubious assumption, in turn, is based on the even more dubious, but extremely powerful, claim that nature is simple.[1] As it turns out, the assumption of the simplicity of nature is at the root of most of our intuitions about the aesthetics of scientific explanation. We would not be concerned with the economy, elegance or symmetry of explanations unless we assumed that nature was simple. However, in so doing we are mistaken, Mill, Peirce, and Russell *et al.* notwithstanding.[2] The mistake derives from employing a correspondence principle of sorts which confuses characteristics of the ontology of nature with epistemological considerations.[3] Thus, it is assumed that the alleged simplicity of nature can be known in an equally simple manner. It is this assumption and its effect on our conception of explanation which is under examination here. I begin with a look at some considerations which suggest that the use of the assumption of simplicity is misguided when considering the adequacy of an explanation, and conclude with an example drawn from the Copernican Revolution.

I. Simplicity

Unfortunately, for the aesthetics of scientific explanation, while at some level nature may be simple, our *understanding* of nature is not. To see this we need only consider the following (1) our understanding of nature proceeds through the development of theories which tackle nature at a variety of different levels: macroscopic, microscopic, submicroscopic, and so on. (2) Even if we were to assume that these levels were hierarchically arranged, simplicity, in any significant sense, could only be achieved by invoking a principle of reduction. But (3) even with a principle of reduction to order our theories, and allowing the assumption that, at some fundamental level, nature may be ontologically simple, it does not follow that our *explanations* of nature can be rendered simple at any level. Even where the concern is located in the fundamental ontology of nature, our explanations are usually of events and processes that take place at "higher," and, hence, more complicated levels of activity. The more complicated the domain of inquiry, the less simple the explanation.

What is being suggested flies in the face of conventional wisdom. The standard DN account of explanation is, as we are often been told, the very model of simplicity: a deductive argument with at least one general premiss.[4] But, the theory of explanation which gives the DN model its force assumes the very simplicity which the model exhibits. Furthermore, that same question-begging assumption is the source of many of the criticisms of the DN model. Consider, for example, the argument presented by Eberle, Kaplan and Montague, which shows that, given Hempel's motivation regarding fundamental theories and laws, which is simplicity as exhibited by generality, then any fundamental theory explains any fact, which is surely an undesirable conclusion. The way around this problem, as provided by Kaplan and separately by Kim, is to restrict the generality, which reduces the simplicity of the conception in both theory and fact.[5]

The assumption of simplicity is again rejected when the deductive-nomological theory of explanation is forced to acknowledge probabilistic explanation. Here, however, the solution to the problem of how to deal with non-deductive connections cannot be handled through a formal maneuver. As Salmon has so clearly argued, a full account of statistical and probabilistic explanation requires more than a logical fix; it requires an account of causality, and providing an account of causality is no simple matter.[6]

Earlier it was claimed that complicated domains require more complicated explanations than do simple domains. This is true even on the DN account. Despite the fact that the structure of each DN explanation is the same, with its general law and statements of initial conditions, why that structure should be accepted as providing an explanation can only itself be explained if we realize that (1) the purported explanation represents a cognitive partition and (2) our understanding of the explanation entails that (a) we know it represents a deliberate isolation of some discrete part of a larger domain and, furthermore, (b) we must have some understanding of both the scope of the domain and what kind of a partition of that domain this partition represents. In other words, the deduction of an observation statement from an *explanans* cannot be an explanation unless we understand the context from which the *explanans* is drawn. An understanding of the context of the explanation requires being able to explain why that particular *explanans* is appropriate to this problem. The more complicated the context, the more complicated the explanation of that *explanans*. That is, one needs not only to justify the selection of that *explanans*, but also, for example, to explain why alternatives were ruled out. Thus, the simplicity of the deductive form of a DN explanation is an illusion. For its cognitive success rests on its being embedded in a larger cognitive structure to which we have access and over which we have control.

II. The Complexity of Cognitive Contexts

The complications suggested here increase once we realize how cognitive contexts themselves get complicated. Consider for a moment an attempt to explain eye color. Eye color is a phenomenon observed at the macro level. That is, it is something you and I see in the ordinary way in ordinary circumstances. At some point in our explanation we retreat from the macroscopic to the microscopic level and start talking about the expression of genes, but a full account of that concept requires further submersion into the depths of micro analysis in order to discuss DNA. Furthermore, this is not a reductive analysis. However, it does appeal to certain theories of chemistry, evolution, and physics, among others. The point here is to provide a context within which to understand the phenomenon of eye color, not to produce a deduction. Such understanding operates at several levels simultaneously, shifting back and forth, crossing lines and causal hierarchies. It requires a series of theories which, while not obviously related, for example, theories of light and theories of chemical interaction, are nevertheless capable of being brought to bear on a single project. To muster these resources requires (a) an understanding of the scope of the problem, in the process of which the partition mentioned earlier is created, (b) the range of tools available, which requires knowledge of the cognitive structure as a whole or some significant part thereof, and (c) the way in which the various levels of the structure relate to each other, both internally and hierarchically. Finally, this kind of explanation is not a deduction, but a *story*. Only stories can capture the complexity of natural phenomena. The narrative format has the flexibility to handle the richness of the process which a DN account does not. To approach such a complicated issue as eye color using DN we would have to provide an artificial structure of stacked or nested DN explanations, and even then the rationale for the explanatory story would probably not be obvious.[7]

In short, while one example does not conclusively make the case, it is at least clear in what ways scientific explanations become complicated and what is needed to handle them. The goal of such an explanation is completeness not simplicity.

But, it might be objected, granting the complexity of the issue, it does not follow that the story has to be complicated. Surely, the objection continues, even the most complicated case can be made simple if broken down into small even pieces. But to take this line confuses comprehensibility with simplicity. A simple tale cannot fully explain a problem such as eye color. But however complex the story, a skillful narrator can usually bring understanding to most audiences. The issue, as I see it, is not to make the phenomenon simple, but to make it comprehensible. To insist on the simplicity of the phenomenon in the light of its genuine complexity is misleading and no explanation at all.[8]

Finally, there is the question of what we actually mean by "simple." There have been many analyses of the concept, some more fruitful than others.[9] But they have one thing in common. Despite the fact that simplicity is fundamentally a comparative notion, these concepts while comparative are all linear. That is, they produce a result which allows for comparison of theories, explanations, computations, and so on, *at the same level only*. Thus, when comparing the Copernican and Ptolemaic systems of astronomy Kuhn notes that Copernicus' system was not simpler than Ptolemy's because they both used roughly the same number of epicycles and eccentrics.[10] Other forms of comparison have to do with the number of steps in a proof, the number of primitive concepts in a theory and the number of auxiliary hypotheses needed in an explanation. What these accounts lack is a means of comparing explanations which are rich in the complex manner alluded to here. Furthermore, if this characterization of the necessary complexity of scientific explanations is correct, then it may not be possible for such comparisons to be made in a helpful fashion.[11] Let us look at the Copernican example in a bit more depth in order to see the kind of problem that makes nonsense of the use of simplicity as a criterion in decision contexts and hence for our account of explanation.

III. The Role of Simplicity in Incomplete Contexts

A rational agent asked to choose between Ptolemy and Copernicus would have to choose Ptolemy. Since Galileo apparently defended Copernicus, that makes Galileo irrational. Now since Galileo, like our other heroes, cannot be irrational, something has gone wrong here. The culprit is the manner in which the issue is posed. The situation here is difficult, even for the most rational of agents. To begin with, the choice is not between two fully developed rivals, nor is it between two equally undeveloped alternatives. It is between a well developed and long entrenched astronomical theory with very strong conceptual ties to a major conceptual system on the one hand, and a mathematical astronomical system with little or no outside conceptual support on the other. For, at the beginning of the seventeenth century, Copernicus' theory constitutes at best a suggestion around which an alternative conceptual system might be constructed at some later point. No agent can be expected to make a fully rational choice between a full scale conceptual system and the mere suggestion of an alternative, at least not on standard accounts of rationality.[12]

Ptolemy's astronomy, while not exactly in harmony with the Aristotelian framework, could, nevertheless rely on a good deal of its metaphysics and physics for support. Copernicus' astronomical system, on the other hand, stood alone without a physics to explain its organization, or a metaphysics to provide credibility for its cosmology. As Copernicus himself claimed, it is a mathematical system. Now, to argue that such an incomplete account ought to be preferred to another which is well-embedded in a reasonably complete world-view seems odd. Furthermore, to argue for it on the grounds of simplicity seems irrational. But this is exactly what Galileo did. Basically his reason for preferring Copernicus was that, on his assumption of the motions of the earth, Galileo could produce a solution to a problem which he believed had not been properly resolved, to wit, the problem of the tides. Galileo's solution to the tides problem cannot be developed here.[13] Nevertheless, several things important for present purposes can be identified. (1) Galileo's selection of Copernicus did not provide additional support for his major scientific work. (2) Galileo chose the simpler account only in the sense that it did not bring any unwanted baggage with it. But (3), without the baggage, Copernicus' account could not be judged more simple than Ptolemy's because it was incomplete. Let us briefly look at each of these points.

The most puzzling aspect of Galileo's defense of Copernicus concerns the fact that Copernicus' astronomical theory added no support to Galileo's new science of motion. As best it could provide a context for his telescopic observations, but to look at it this way is to proceed backwards. If anything, Galileo's observations added support for Copernicus, not the other way around, except when it came to Galileo's account of the tides.

Let us consider the second and third items together. It is true that Copernicus' astronomy was minimally simpler than Ptolemy's *in this sense only*: as an account of how to compute motions of the planets. It did not have the means to explain how the planets moved, what they were composed of, their dynamical relations one to another, and so on. Ptolemy's account, on the other hand, was, however incorrectly, linked with Aristotelian theories of motion and matter to explain how it was that things moved as they did. And, granted, while there were problems reconciling Ptolemaic motions and Aristotelian quintessence, at least the Aristotelians had an explanation, if a bad one. Copernicus had none. And to choose a theory with no explanation over one with a bad explanation, is, in a world inhabited only by philosophers, to choose the simpler.

But things are not as bad as they seem. Galileo did have *a reason* for preferring Copernicus over Ptolemy in addition to the fact that Copernicus' account helped him out in his efforts to solve the problem of the tides: Galileo was opposed to the entire Aristotelian explanatory method. Thus, to the extent that he could provide an explanation of the tides without having to deal with the Aristotelian appeal to first principles, he felt he had a simpler account. And so Galileo did choose Copernicus' theory because it was simpler, but that same sense of simplicity also ultimately rendered his explanation of the tides incomplete because Copernicus' system was so skeletal.

And this leaves us with the following view of the situation. Galileo selected a small theory to help him solve an isolated problem. But he failed to explain the tides because no alternative theory of celestial motion compatible with Copernicus' account was available to supplement the theory in the way that Aristotle's physics supplemented Ptolemy's. Without Aristotelian physics Galileo could not explain why the tides moved as they did even in a Copernican universe. And so he cheated. In the end he relied on an Aristotelian account of matter, the heaviness of water, and uniform circular motion.

Thus, despite the appeal of Copernicus' theory on grounds of simplicity, in his efforts to provide an explanation of the tides, Galileo was forced to look beyond simplicity and fill out his account in order to make it comprehensible. Although I have not elaborated the details of Galileo's explanation, it is safe to say the the incompleteness of the explanation of the tides was a result of the lack of depth in Galileo's cognitive partition—that is, he needed a theory of motion and of matter in order to make his explanation work. These theories functioned at different levels than the observational level of the tides. Thus, despite the drive for simplicity, Galileo implicitly acknowledges the complexity of the situation and falls back on an *ad hoc* solution.

The conclusion to be drawn from the example is, and again it is important to recognize that one example doesn't clinch the case, the following. While simplicity may sometimes be a factor in theory selection, when it comes to explanation, completeness may play a more significant overriding role. Therefore, simplicity, as a characteristic of nature, is to be distinguished from completeness and comprehensibility as criteria of adequacy for explanations. Furthermore, the aesthetics of simplicity do not play a role in choosing between competing explanations because it is not the simplicity of the explanation which makes the difference, but it is other factors such as completeness and comprehension. Furthermore, as we saw in the Galileo example, often the choice is not between two equally adequate explanations, *ceteris paribus*. In fact that situation never occurs. If there are two competing explanations, prior to completion of a "final" science, they will always be incomplete—often in different degrees. Simplicity, if it has a role to play at all, may be a factor in ontology, but even that is not clear. That is a topic for another time.[14]

Virginia Polytechnic Institute and State University

NOTES

1. Consider, for example, Newton's claim in the first edition of the *Principia*, "nature is simple and does not luxuriate in superfluous causes of things."

2. Russell, Bertrand, *Our Knowledge of the External World* (New York: Norton, 1929); Peirce, C. S., *The Collected Works of Charles Saunders Peirce* (Cambridge: Harvard University Press, 1960); and Mill, J. S., "An Examination of Sir William Hamilton's Philosophy" in *Collected Works of J. S. Mill*, vol. 9 (Torronto, University of Toronto Press, 1963).

3. See Ariew, Roger, "Did Ockham Use His Razor," *Franciscan Studies*, vol. 37 (1977), pp. 5-17, for a nice brief history of the metaphysical versus the methodological use of the principle of simplicity.

4. For what is generally accepted as the classical version of the DN account, see Hempel and Oppenheim, "Studies in the Logic of Explanation," *Philosophy of Science*, vol. 15 (1948), pp. 135-75. Reprinted in *Theories of Explanation*, ed. by J. C. Pitt (Oxford: Oxford University Press, 1988).

5. Eberle, R., Kaplan, D., and Montague, R., "Hempel and Oppenheim on Explanation," *Philosophy of Science*, vol. 28 (1961), pp. 418-28. Kaplan, D., "Explanation Revisited," *Philosophy of Science*, vol. 28 (1961), pp. 429-36. Kim, J., "Discussion: On the Logical Conditions of Deductive Explanation," *Philosophy of Science*, vol. 30 (1963), pp. 286-91.

6. Salmon, W. C., *Scientific Explanation and the Causal Structure of the World* (Princeton: Princeton University Press, 1984).

7. For the notion of a narrative I am indebted to Wilfrid Sellars' account of a theory as a logical core with a commentary. See Sellars, W., "Empiricism and the Philosophy of Mind," in *Science, Perception, and Reality* (London: Routledge and Kegan Paul, 1963).

8. In part, this account of simplicity versus comprehension is an attempt to respond to those who maintain that any subject can be explained in terms simple enough for anyone to understand. There is a certain sense in which that is true—that is, a skillful narrator should be able to explain the complexity of a situation to just about anyone. What it does not mean is that every subject matter is inherently simple and we, philosophers and

scientists, just make it complicated to impress people.

9. For examples of good work see Elliott Sober, *Simplicity*, Oxford, 1975; Mario Bunge, "The Complexity of Simplicity," *The Journal of Philosophy*, vol. 59 (1962), pp. 113-34.

10. Kuhn, Thomas, *The Copernican Revolution* (Cambridge, Mass.: Harvard University Press, 1957), pp. 169-71.

11. Furthermore, the point of developing a conception of simplicity which allows for such comparisons has been to develop a criterion for choosing among theories and explanations. In that program simplicity is usually one among a number of such concepts. Now the whole point of developing such a concept may well be in doubt if the project of developing a theory of rational choice for selecting among competing theories, hypotheses, explanations, etc., is itself in doubt. While there are many good reasons to question the general program, they cannot be pursued here. Suffice it to say that most such theories leave out the role of technology in the development of science and hence conceive of the decision-making process in a highly artificial context.

12. For an alternative account under which the choice can be seen to be rational, see Pitt, J. C., "The Untrodden Road: Rationality and Galileo's Theory of the Tides," *Nature and System*, vol. 4 (1982), pp. 87-99.

13. See Pitt, J. C., "The Untrodden Road: Rationality and Galileo's Theory of the Tides," *Nature and System*, vol. 4 (1982), 87-99; and "Galileo's Conception of Rationality," *Philosophy of Science*, vol. 55 (1988), pp. 87-103.

14. I wish to thank Roger Ariew and J. O. Falkinham, III, for their help and careful criticism, I obviously did not heed it all.

SIMPLICITY AS AN EPISTEMIC VIRTUE: THE VIEW FROM THE NEURONAL LEVEL

Paul M. Churchland

MY aim in this short paper is twofold. First, I will try to illustrate the poverty of the idea that one can usefully divide the epistemic virtues of a theory or hypothesis into two disjoint classes: the empirical and the superempirical. Second, I will argue that traditional attempts to capture the dimensions of epistemic virtue at the syntactic or semantic levels are ill-conceived, and should be superseded by a more strongly empirical approach that focuses instead on the representational and computational strategies used in the micro-architecture of the biological brain. At that level, as I shall show, one can define a straightforward measure of the *simplicity* of a class of representations, and even see why it is a genuinely epistemic virtue, as opposed to a "merely pragmatic" virtue. The general conclusion to be drawn is that traditional epistemology has chronically focused its efforts at the wrong level of analysis, and that attention to this deeper level of neural computation may reopen the doors to epistemological progress.

There is little or no consensus as to just what the list of superempirical virtues (SEVs) should contain. But simplicity would be on almost everyone's list, and it will be the focus of attention in what follows. This important notion has two problems: it is robustly resistant to attempts to define or to measure it, and it is not clear why it should be counted relevant to judgments of truth and falsity in any case. There seems no obvious reason, either *a priori* or *a posteriori*, why the world should be simple rather than complicated, and epistemic decisions based on the contrary assumption thus appear arbitrary and unjustified.

With the empirical virtues (EVs), the situation is supposed to be quite different. Sentences known true by observation may still underdetermine one's general hypotheses, but consistency with known truth is at least a clearly truth-*relevant* consideration. There seems nothing comparably arbitrary about preferring hypotheses that are consistent with the empirical facts over hypotheses that are not.

This is surely true, unless there is some hidden arbitrariness in the choice of what counts as the "empirical facts." And of course, there *is*. Any empirical description or empirical judgment must deploy some categorial framework or other, and such frameworks are always speculative *theories* as to some of the classes into which Nature divides herself, and some of the general relations that hold between them. Most importantly, the intrinsic character of our raw sensory experience does not determine a uniquely appropriate categorial framework. Indefinitely many quite different frameworks might be used to variously express what will be called the "empirical facts" by their users, and those "facts" will vary substantially from framework to framework. If we are to distinguish the "true" categorial framework from the teeming multitude of pretenders, we must ask which categorial framework allows us to formulate the *simplest overall set of laws* connecting its proprietary observational categories. This will be, roughly, the set of laws that is minimally plagued by sundry qualifications, exceptions, and outright failures.

Here we have considerations of simplicity emerging as a major determinant in our choice of an observation framework within which to express our descriptions of the "empirical facts." The general lesson is that we do not have access to any facts, not even the "empirical" facts, that is independent of considerations of simplicity. Accordingly, the contrast standardly drawn between the purely empirical virtues on the one hand, and the superempirical virtues on the other, is deeply misconceived. Considerations of simplicity are a major factor in epistemic decisions at all levels, including the decision of which observation vocabulary to accept in the first place.

One casualty of this result is the specific sort of anti-realism advanced by van Fraassen,[1] the sort which requires of us an agnostic attitude towards *un*observable ontologies but allows a realistic attitude towards an ontology of observables. What motivates van Fraassen's anti-realism is 1) the underdetermination of theory by the EVs (briefly, many quite different models will share a common empirical substructure), and 2) the merely pragmatic relevance of the SEVs, which means that the ambiguities left open by the purely EVs cannot be relevantly closed by the SEVs. What is thus left open cannot merit our belief, says van Fraassen. But what is not thus left open (the common empirical substructure) can.

What is wrong here is the privileged and contrastive position ascribed to the so-called empirical or observational substructures of the models at issue. It may well be that our theoretical beliefs are imperiled by an underdetermination that is not adequately compensated by considerations of simplicity. But the very same is true of our observation beliefs, and for our ascriptions of observability, for they also are inescapably hostage to theory, and hence to problematic considerations of simplicity. If the SEVs pose a problem (and surely they do), they pose a problem for cognition generally, not for just a part of it. That is why positivistic instrumentalism and constructive empiricism are not appropriate responses to the problems we face.

What would be a more appropriate response? This question has an edge to it, since we seem to have fallen from the frying pan into the fire. The credibility of unobservables no longer exhaust our worries; it is the credibility of our entire ontology that is thrown into doubt. And SEVs such as simplicity are still at the center of the problem.

There are any number of things one might try at this point. I propose to see if the situation looks any more tractable when examined from the perspective of the "microstructure" of cognition: that is, from the perspective of some recent results in cognitive neurobiology and connectionist AI.

It now appears that a major technique, perhaps *the* major technique, for representation in the brain is *vector coding* of complex states of affairs. This requires a number, often a very large number, of parallel nerve fibers, each of which is conveying a train of tiny pulses (action potentials or "spikes") with a specific frequency, from the coding site to some other place in the brain. The momentary pattern or profile of spiking frequencies across that set of parallel fibers is the coding vector. (Think of it as just an ordered set of numbers, each one representing the frequency in one of the fibers in the collective pathway.) Depending on the nature of the stimulus at the coding site, the coding vector sent down the collective pathway will vary. The idea is that each distinct type of initial stimulus receives a unique coding vector as its representation within the brain. In this way can the brain code the various colors, smells, tastes, sounds, faces, and so forth that it encounters. It can also code motor output vectors in this same way, and thus exercise coherent control over the complex system of bodily muscles.

The abstract virtues of vector coding are discussed elsewhere,[2,3] so I shall not pursue them here, save to mention just one. Vector coding is also part of a highly general solution to the problem of how the brain *computes*. The brain can perform very powerful computations over these representations by executing sundry vector-to-vector transformations, as displayed for example in multiplying a vector by a matrix to get a new and different vector (see fig. 1b).

Here the four axons of the incoming horizontal pathway make multiple synaptic connections with the dendritic filaments of the waiting Purkinje cells. The various sizes, or "weights" as they are called, of the several synaptic connections constitute the coefficients of the matrix. The three Purkinje cells sum the axonal activity that reaches them through these many connections, and the three output axons convey an appropriate triple of axonal activation levels away from the matrix. This is the output vector. The brain has a great deal of wetware that is well-suited to computational activities of this general kind. The entire cerebellum, for example, appears to be a kind of matrix multiplier for the very high-dimensional vectors required in motor control (see fig. 1a).

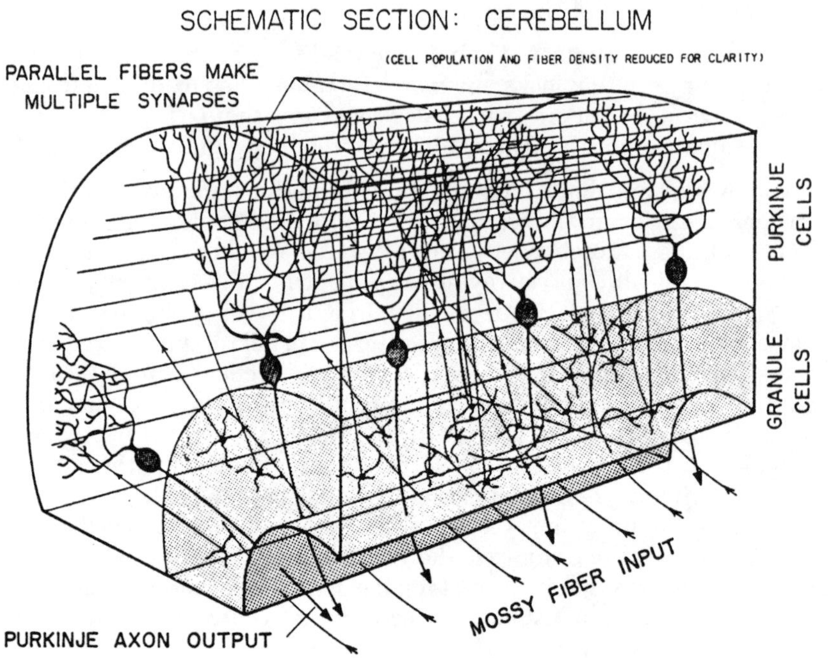

FIGURE 1a

Such transformations need not be limited to simple linear transformations, as the reference to matrix multiplication might suggest. If the transmission properties of the synapses are nonlinear, or if the output function of each Purkinje cell is nonlinear, then a system of the kind portrayed in fig. 1b can execute radically nonlinear transformations, which enlarges dramatically the range of computations it can perform.

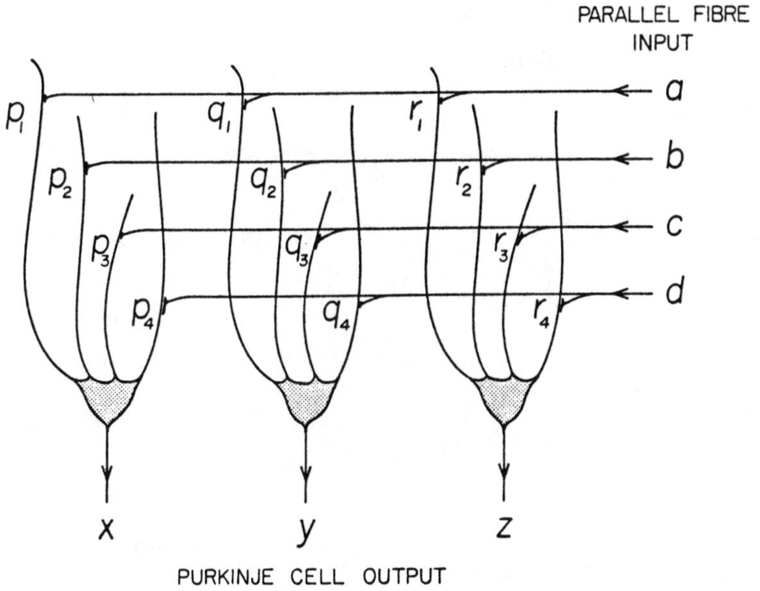

FIGURE 1b

I now direct your attention to a class of *artificial* devices designed to explore the cognitive properties of such vector-transforming systems. They are called "neural nets" or "associative nets," and they are currently the focus of much attention in connectionist AI and cognitive neurobiology. What makes them interesting in the present context is that they can learn (= be trained) to recognize important similarities among diverse and noisy examples of input stimuli and to give a uniform response to elements of the relevant similarity class. The network of fig. 2, for example, can be trained to discriminate the sonar echoes returned from rocks from sonar echoes returned from metallic mines of the same general size.[4]

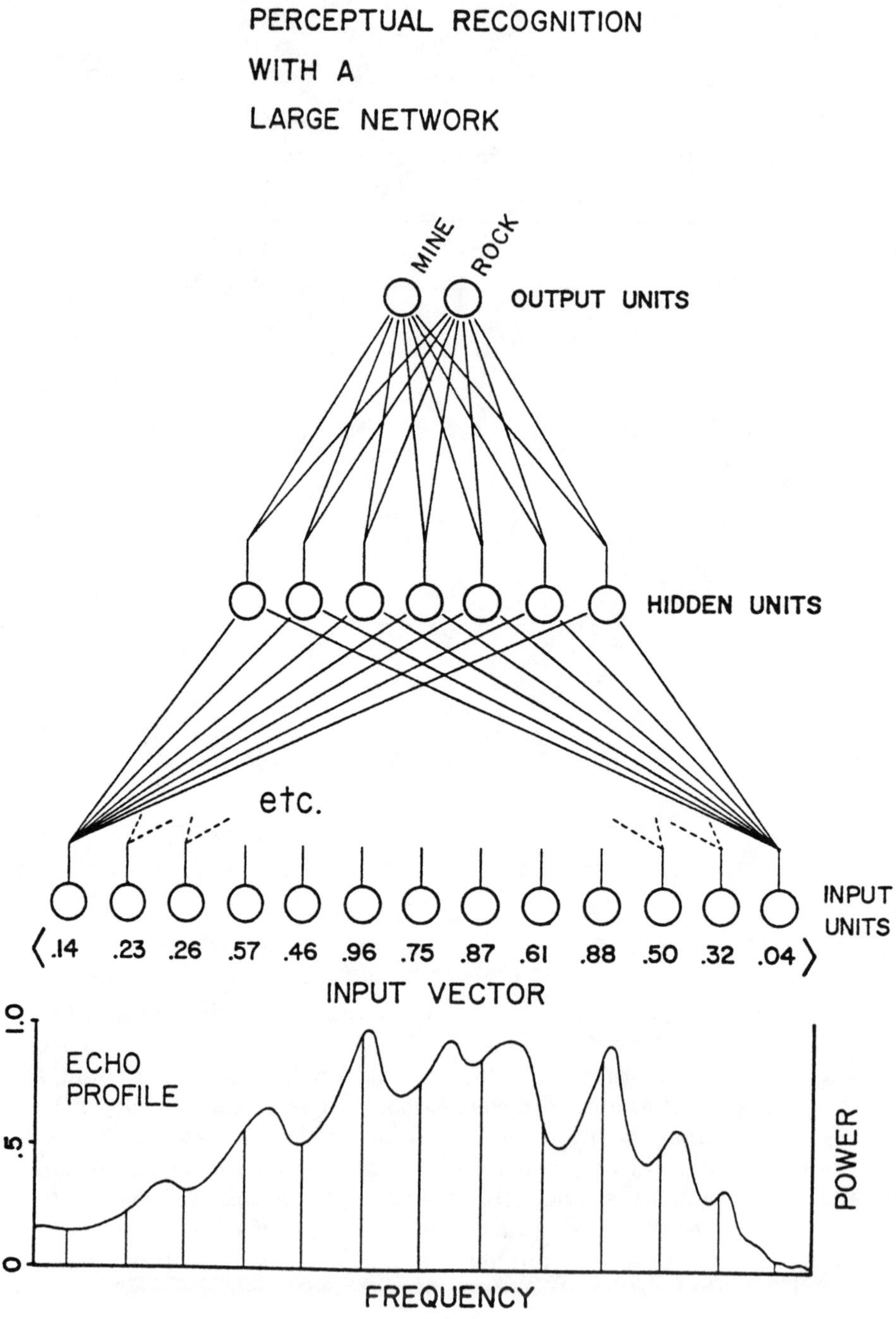
PERCEPTUAL RECOGNITION
WITH A
LARGE NETWORK
MINE
ROCK
OUTPUT UNITS
HIDDEN UNITS
etc.
INPUT UNITS
⟨.14 .23 .26 .57 .46 .96 .75 .87 .61 .88 .50 .32 .04⟩
INPUT VECTOR
ECHO PROFILE
1.0
.5
0
POWER
FREQUENCY

FIGURE 2

Very briefly, it works as follows. A given echo is sampled in a frequency analyzer (not shown) for its total energy levels at various frequencies (as represented at the bottom of fig. 2), and these levels are then entered as elements of the input vector: that is, as activation levels in the input units at the bottom layer. Activation is then propagated upwards through the many connecting fibers, and through their various synaptic connections with the "hidden" units at the middle level, and ultimately to the output units at the top. The activation pattern is profoundly transformed in the process, and it finally produces an output vector in the form of a pair of activation levels across the two output units. An appropriately trained network will produce a common output vector (<1, 0>, say) for any mine echo entered as input. And it will produce a common output vector (<0, 1>, say) for any rock echo entered as input, even though the two types of echo are indistinguishable to the casual ear.

We teach it this somewhat surprising skill by 1) feeding it a long series of diverse examples of each type of sonar echo, 2) examining the (usually incorrect) output vector it produces in each case, and then 3) adjusting the many synaptic weights according to a rule[5] that steadily minimizes the difference between the output vectors the network actually produces and the output vectors it should be producing to discriminate the two kinds of echoes successfully. This training procedure is standardly executed by an auxiliary computer programmed to feed sample echoes from a "training set" to the network, monitor its responses, and adjust its weights according to the special rule after each trial. Under the pressure of such repeated corrections, the behavior of the network slowly converges on the behavior we desire (see fig. 3). It gives something close to a <1, 0> output when it hears a mine echo from the training set; and it gives something close to a <0, 1> when it hears a rock echo from the training set. It has reached a configuration of synaptic weights that now responds appropriately to each and all of the various inputs in the training set. It now has a successful internal representation or *theory* to guide its behavior. That theory is represented, or better, it is constituted, by a specific configuration of wynaptic weights at the level of the hidden units and at the level of the output units.

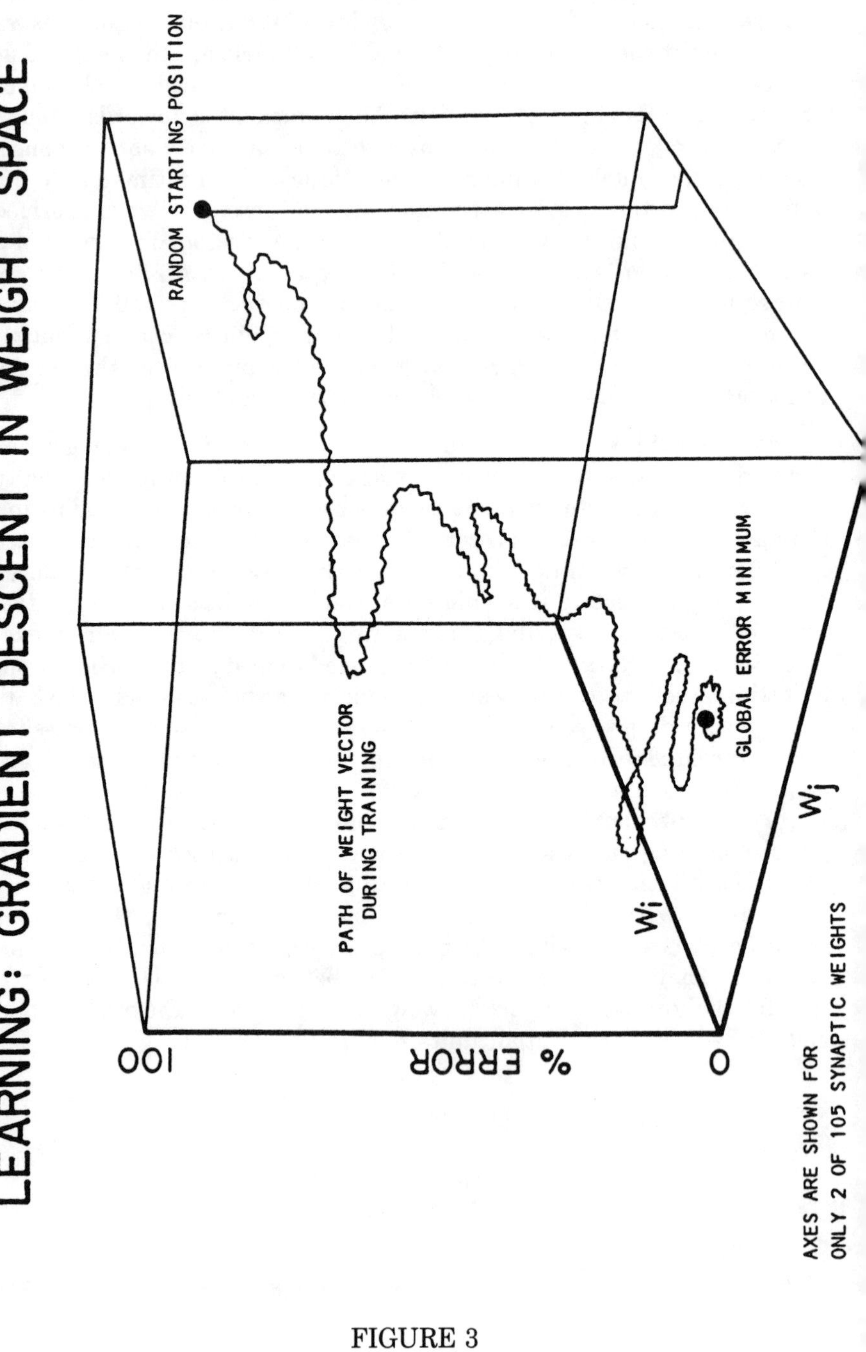
LEARNING: GRADIENT DESCENT IN WEIGHT SPACE
RANDOM STARTING POSITION
PATH OF WEIGHT VECTOR
DURING TRAINING
GLOBAL ERROR MINIMUM
100
% ERROR
0
W_i
W_j
AXES ARE SHOWN FOR
ONLY 2 OF 105 SYNAPTIC WEIGHTS

FIGURE 3

Now, what happens when we enter into the network a vector for a *novel* mine echo, one from outside the set of samples on which it was trained? The network depicted in fig. 2 responds surprisingly well to new mine echoes, even though they may differ substantially from those in the training set. This is because the network, in the course of training, has succeeded in finding a mode of representing the various mine echoes, at the level of the hidden units, that "sees past" their superficial variety so as to code them in a more-or-less uniform way, a way that the units at the output level can discriminate in turn and thus respond with the desired <1, 0> output vector (see fig. 4). If the system has succeeded in finding such a successful coding strategy, all that is required for successful generalization is that the new mine echoes *resemble* the mine-echoes in the training set in the same general sorts of ways in which those in the training set resemble each other.

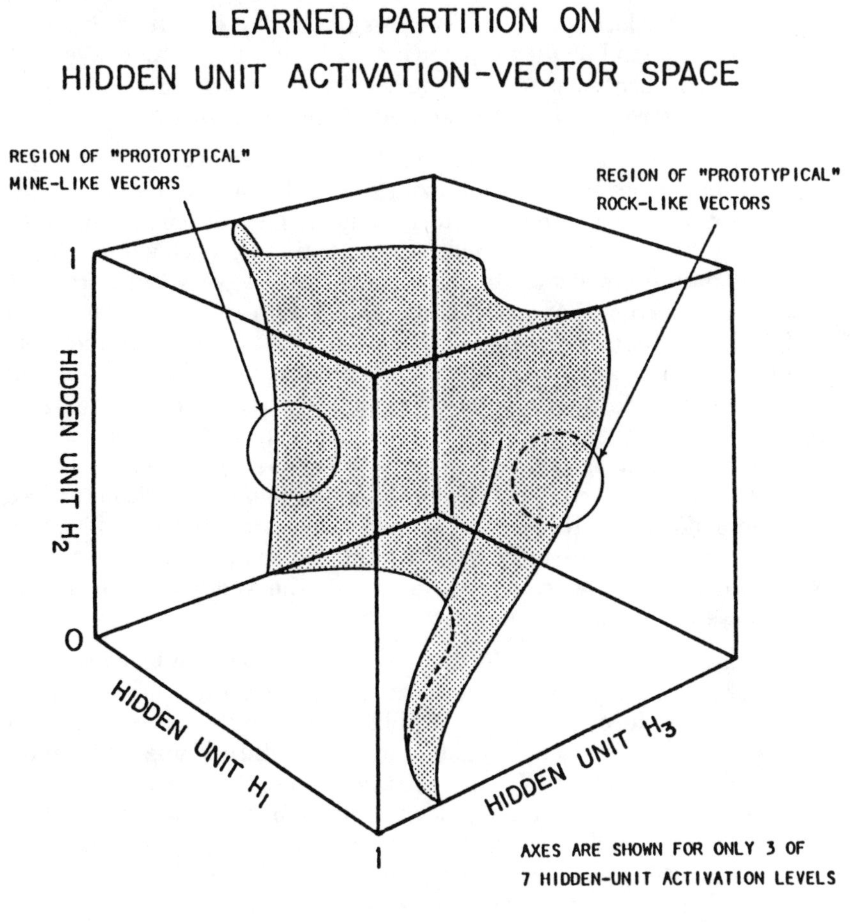

FIGURE 4

Networks of the kind at issue can thus generalize beyond the training set very successfully indeed. But they do not always do so, and it is interesting what determines their success in this regard. How well the training generalizes is a function of *how many* hidden units the system possesses. There is, it turns out, an optimal number of hidden units for any given problem. With more than the optimal number, the system learns to respond appropriately to all of the various samples in the training set, but it generalizes to new samples only very poorly. With less than the optimal number, it never really learns to respond appropriately to all of the samples from its training set.

The reason is as follows. What the network is doing during the training period is gradually to generate a set of internal representations at the level of the hidden units. One class of hidden-unit activation vectors is characteristic of rock-like input vectors; another class is characteristic of mine-like input vectors. The system is *theorizing* at the level of the hidden units, exploring the space of possible coding vectors, in hopes of finding some partial or global partition on it that the output layer can then exploit in turn, so as to draw the desired distinctions and thus bring the process of error-induced synaptic adjustments to an end.

If there are far too many hidden units, then the learning process can be partially subverted in the following way. The lazy system cheats: it learns a set of *unrelated* representations at the level of the hidden units. It learns a distinct but unrelated representation for each sample input (or for a small group of such inputs) drawn from the very finite training set, a representation that does prompt the correct response at the output level. But since there is nothing common to all of the hidden-unit rock-representations, or to all of the hidden-unit mine-representations, an input vector from outside the training set produces a response that bears no relation to the representations already formed. The system has not learned to see *what is common* within each of the two stimulus classes, which would allow it to generalize effortlessly to new cases that shared that common feature. It has just knocked together a "look-up table" that allows it to deal successfully with the limited samples in the training set, at which point the error messages cease, the weights stop evolving, and the system stops learning.

There are two ways to avoid this ad hoc, unprojectible learning. One is to enlarge dramatically the size of the training set. But a more effective way is just to reduce the number of hidden units in the network, so that it lacks the resources to cobble together such wasteful and ungeneralizable internal representations. We must reduce them to the point where it has to find a *single* partition on the hidden-unit vector space, a partition that puts all of the sample rock-representations on one side, and all of the sample mine-representations on the other (see again fig. 4). A system constrained in this way will generalize far better, for the global partition it has been forced to find corresponds to something *common* to each member of the relevant stimulus class. In the network at issue, the final partition drawn in hidden-unit vector space corresponds empirically to the distinction between metallic and non-metallic objects. This is modestly amazing: under the steady pressure of error-correction, the network has found an underlying uniformity behind the sensory diversity of its input vectors.

On the other hand, if the network has too few hidden units, then it lacks the resources even to express a coding vector that is adequate to characterize the underlying uniformity, and it will not master completely even the samples from the training set. In other words, simplicity may be a virtue, but one must command sufficient complexity to meet the task at hand.

We have just seen how forcing a system to generate a smaller number of more economical representations or coding vectors at the level of the hidden units can produce a system whose learning achievements generalize more effectively to novel situations. *Ceteris paribus*, the simpler hypotheses generalize better. Getting by with fewer resources is of course a virtue in itself, though a "pragmatic" one, to be sure. But that is not the principal virtue displayed. Superior generalization is a genuinely epistemic virtue, as opposed to a "merely pragmatic" virtue, and it is regularly displayed by networks that are constrained, in the fashion described, to find the simplest hypothesis concerning whatever structures might be hidden in or behind its input vectors.

Of course nothing *guarantees* successful generalization: a network is always hostage to the quality of its training set relative to the total population. And there may be equally simple alternative hypotheses that generalize differentially well. But from the perspective of the relevant microdynamics, we can see at least one clear reason why simplicity is more than a merely pragmatic virtue. It is an epistemic virtue, not principally because simple hypotheses avoid the vice of being complex, but because they also avoid the vice of being *ad hoc*.

As we have seen, a sub-linguistic or neurocomputational perspective provides us with powerful insights concerning the possible nature of mental representations, sensory recognition, concept learning, and simplicity. And it does so in a unified fashion. This approach to epistemology clearly merits further exploration.

I close with a disclaimer. The concept of simplicity employed here clearly does not do justice to the complex notion we use at the linguistic level and in the philosophical tradition. But this does not trouble me overmuch, because I think that the simplicity pursued by philosophers of science is rather like the common cold: it is many different things masquerading as a single mysterious thing. In order to separate out the important dimensions and elements of theoretical simplicity, we need to reconstruct the issues within a new theoretical framework: specifically, the framework of cognitive neurobiology and connectionist AI. We have already seen some of the insights this will provide. After we have won the battle at this level, then, perhaps, may we return with profit to the traditional level of linguistic representations.

University of California, San Diego

NOTES

1. van Fraassen, B., *The Scientific Image* (New York: Oxford University Press, 1980).

2. Churchland, P. M., "Some Reductive Strategies in Cognitive Neurobiology," *Mind*, vol. 95 (1986), pp. 279-309.

3. Churchland, P. M., *Matter and Consciousness*, 2nd ed. (Cambridge, MA: MIT Press, 1988).

4. Gorman, P., and Sejnowski, T., "Learned Classification of Sonar Targets Using a Massively-Parallel Network," *IEEE Transactions: Acoustics, Speech, and Signal Processing* (forthcoming, 1988).

5. Rumelhart, D., Hinton, G., and Williams, R., "Learning Representations by Back-propagating Errors," *Nature* vol. 323 (1986), pp. 533-36.

TAMING A REGULATIVE PRINCIPLE: FROM KANT TO SCHLICK

Matti Sintonen

KANT'S role as a philosophical canonizer of the Newtonian worldview is now common knowledge. What is less well known is that his first and third *Critiques* contain an account of the role of aesthetic properties which looms behind more modern discussions. Kant was aware and much impressed by progress in the systematization of natural knowledge, and he gave a classic characterization of increase in knowledge as increase in cognitive law and order. But he was equally aware of the pitfalls in the assumption that nature as such is simple and harmonious, and his genius was in his account of the justification of our striving for unification. In this paper I am primarily interested in the justification of unification (and other aesthetic properties) as cognitive virtues, and only secondarily in their more precise nature. I shall try to show that the Kantian justification eroded when the Newtonian worldview itself was undermined, leaving no obvious alternative mode of justification.

I. Kant and the Principle of the Finality of Nature

The ideal for Kant was a systematically organized body of knowledge in which every part is related to other parts, "in conformity with a single principle."[1] Reason's legislative prescriptions require that the totality of our knowledge (concepts and laws) fulfills certain systemic and structural requirements. The key notion is that of the architechtonic of Pure Reason, which refers to certain features of a building or an organism: on one hand our body of knowledge should be complete and contain no gaps (questions raised but not answered by the cognitive faculty); on the other hand it should be harmonious in the sense that all parts of the edifice fit in an "intelligible" pattern prescribed by reason.[2]

Unity manifests on all three levels of the cognitive faculty. On the level of sensibility we have the *a priori* forms of intuition (i.e., absolute space and time) which guarantee perceptual unity of apperception (and, recall, perception is *aisthesis*). On the level of understanding there are the Categories (most notably, Causality) which require that any experiences that allow lawful representation have a certain form. But the most important source of aesthetic considerations, of systematicity, appears on the highest level, as teleological and purposive harmony imposed by Pure Reason.[3]

The problem for Kant clearly was the contingent, compartmentalized, and purely mechanical nature of empirical laws. Since contingency is repulsive and mechanics alien to Reason, it cannot help trying to find (or create) order and harmony, and more specifically, purposive harmony. It is the task of judgment to reduce this underdetermination, and to show that laws fall into an intelligible pattern. To this end judgment presupposes that the world is *as if* designed for our cognitive faculties. This is anthropomorhism, but according to Kant unavoidable and, in fact, the only path to understanding—or at least, human understanding.[4]

Given that the goal of human knowledge is a maximally systematic representation of nature, the next question is why this is the goal for us, that is, what makes *us* search for such a representation. It is to Kant's credit that he did not dismiss this question as meaningless. And he did have an answer, in the *Critique of Judgment*, which rounds off the entire critical enterprise.[5] The first *Critique* examines the presuppositions of the faculty of understanding, that is, its *a priori* ingredients, and the second those of desire, postulating the notions of God, freedom and immortality. It is the task of judgment to serve as a "middle term" between understanding and reason by providing an *a priori* rule for pleasure. Pleasure is according to Kant tied to human motivation and desire, but Kant also maintains that pleasure marks the transition from the faculty of understanding (pure knowledge) to the concept of freedom.[6]

No pleasure accompanies the falling of perceptions under the *a priori* forms of intuition, or of the laws under Categories, because here understanding follows its own natural inclinations. But that "two or more empirical heterogeneous laws of nature are allied under one principle that embraces them both, is the ground of a very appreciable pleasure, often even of admiration . . . ."[7] Kant has, then, found a motive in the pleasure which we feel when seemingly heterogeneous laws conspire in accordance with the architechtonic wishes of reason, "in this their accord with our cognitive faculty." On the other hand, lack of accord is accompanied by displeasure, a fact which provides an incentive to attempt to complete reason's grand design.

The aesthetic properties of unity, harmony and order are dictates of specifically human reason, and hence subjective. Yet they are not biased or dependent on this or that individual. For Kant aesthetic judgments in general are not mere statements of personal preference. Although based on a feeling of pleasure, their claims are something which command intersubjective agreement. We are entitled, he thinks, to expect that all other (enlightened) judging subjects, when representing the object to themselves, make the same judgments and feel the same pleasure in the judgment.[8]

What about the justification of the use of the architechtonic design in arguments for a particular judgment in the cognitive area? What guarantees that what pleases the mind is true (or acceptable, or sound)? Nothing whatsoever. That a particular conjecture fits in the desired pattern is a presupposition which judgment takes for granted—it cannot help trying to reduce arbitrariness by relating empirical laws as Pure Reason requires. But the mind can survive drawbacks when the presupposition is not borne out. The presupposition is so "indeterminate" on the question of the extent of fit between nature and our cognitive faculties that if we are forced to see irreducible multiplicity of laws, "we can reconcile ourselves to the thought." But Kant does observe, as did Bacon earlier, that we listen more gladly to others who entertain the hope that, in the process of unraveling "the secrets of nature," ever greater simplicity and harmony is found "in the apparent heterogeneity of its empirical laws."[9]

The idea of unification and harmony derives from the unavoidable tendency of the human mind to see purpose and teleology in nature: we, in virtue of our specifically human form of intelligence, look upon nature as if it were created by an intelligent mind. But this tendency is a regulative, not constitutive, idea, which means that it serves as a heuristic and not as a guarantee of foreseen results. In any case when Nature cooperates, our pleasure is of the disinterested variety, and therefore counts, according to Kant's definition in the third *Critique*, as aesthetic.

II. Schlick and the Claim of Reason

Moritz Schlick, the leader of Viennese Positivism, is one link between Kant and present-day pragmatized theory aesthetics.[10] He inherited from Kant the view that although science is rational pursuit of knowledge and understanding, it is not trivial pursuit. Understanding proceeds step-wise in the rediscovery of something in another thing (that is, through reduction), and in this process "the number of phenomena explained by one and the same principle becomes ever greater, and hence the number of principles needed to explain the totality of phenomena becomes even smaller." The highest level of knowledge is the level "with the fewest explanatory principles that are not themselves susceptible of further explanation." Thus the ultimate goal of all knowledge is to make this minimum as small as possible.[11]

But what is the epistemological or methodological status of this drive to unification? Schlick dismissed the view that we choose simpler theories because nature as such is simple. The view is simplistic because it is always possible to conjure up a complex theory which has the same results as a simpler one—simplicity considerations alone cannot single out a theory as the uniquely correct one. The upshot is that although we are "animated by the belief in the simplicity of nature," this belief cannot be based on experience, but is, rather, "a presupposition that we bring to experience."[12] Another way defines reality anew, so that the simplest set of assumptions is guaranteed to be the one which corresponds to reality. Schlick, however, in accordance with his realist intuitions, appeared to think that the real is an ultimate and admits of no constricting definition. Then, he says, there is but one logical justification for it, namely, that in very many cases "the greater simplicity of a theory depends on its containing fewer arbitrary elements (where "simplicity" does indeed consist in other factors, any logical vindication breaks down). The more hypotheses I introduce to explain a situation, the more assumptions I make, the more different ways there are for securing agreement with experience." In holding to this justification we are "sure of diverging from reality at least no further than is necessitated by the bounds of our knowledge as such."[13]

Clearly for Schlick the commitment to the goal of unification and to the reduction of the arbitrary is a regulative principle in Kant's sense, in that it imputes to nature as much order as it allows—and no more. But the principle obtains in Schlick's hands a new naturalistic twist. He was one of the earliest philosophers to write on the philosophical import of the Principle of Relativity, and his conclusions were decidedly unKantian.[14] Thus when he published his masterpiece *Allgemeine Erkenntnistheorie* (a year later, in 1916) it was no longer possible (for him, anyway) to take Kant's project at fact value. The biggest disaster befell the Kantian justification of absolute space and time as necessary forms of perception—the first level of unification. Schlick's conclusion was, simply, that Kant was misled to attribute to the pure form of (empirical) intuition something which really was contributed by "understanding or reflection."[15] That the downfall of the Newtonian framework had a profound impact on our notion of space and time needs no further documentation. What is interesting is that it contributed to the taming of Kantian Reason too: the idea that the specifically human reason was immutable became suspect.

That Pure Reason had suffered a blow to its universality becomes clear in Schlick's account of the value of knowledge, for knowledge no longer is constrained by the *a priori* principle of "the finality of nature"—that nature must harmonize with our faculty of understanding—but rather by evolution. Knowledge in general contributes towards "the preservation of the individual and the species," and the drive for knowledge undoubtedly falls under this general principle: "In its origin, thinking is only a tool for the self-maintenance of the individual and the species, like eating and drinking, fighting and courting." And Schlick goes on to say that the mechanism of judging and inferring contributes towards better adaptation to the environment than automatic association which focuses on typical cases.

Similarly, his explanation of why the ultimate goal of unification is a goal for us comes from "the province of biology," because pleasure is naturalized. Recall that according to Kant we feel pleasure in the fact that nature succumbs to the wishes of reason, and that, although pleasure belongs to the realm of subjectivity, judgments of beauty still have universal validity. Not so for Schlick. In his early treatise on aesthetics he singles out the question "Why does anything whatsoever appear beautiful?" as the *Grundproblem.*[16] And he immediately goes on to interpret it as a *causal* query: "The 'why' here must be taken to be asking for a real causal explanation; so it is not a matter of specifying the properties in virtue of which an object becomes a beautiful one, but rather of discovering the causes that lead to these properties actually having such an effect."[17]

So, how do we go about explaining why an object is beautiful to us? Kant maintained, in a characterization that set the tone for generations, that aesthetic pleasure is disinterested delight, contemplation from a point of view different from utility. Here Schlick departs from Kant, because he thinks that "the good is good not because it has 'a value in itself,' but because it gives joy." But how does this reply carry over to cognitive or epistemic values? "Why do we strive to designate the rich manifold of the universe by means of only those concepts that are built up from a minimum of elementary concepts?" Schlick does not waver on the answer: "the reduction of one thing to another affords us pleasure . . . . So too the value of knowledge consists quite simply in the fact that we enjoy it."[18]

Agreed that Pure Reason thus becomes naturalized, Schlick still owes us an account of how the naturalized reason has acquired the taste for orderliness. This account runs as follows. Although the direct causes of pleasurable factors are, from a psychological point of view, irreducible, they are not so from an evolutionary point of view, because whatever is useful for an individual or a species must appear to the individual as agreeable. Furthermore, adaptation brings about that what is useful and agreeable betokens pleasure. Thus although an individual acts on the hedonistic principle, the objects of pleasure have an evolutionary explanation. And Schlick notes that once an organism achieves periods when it is not caught in the struggle for existence—when all energy does not go to the satisfaction of the primary needs—all perceptual and ideational activities which once were associated with pleasurable activities in the primary sense come to produce pleasure.

III. Enter Historicism

Let me end this discussion by pointing out some problems for naturalized reason, and by giving some (purely circumstantial) evidence of the influence of Kant and Schlick on present-day theory aesthetics. The gravest problem is that it is not clear why evolution should favor unification, or simplicity *of the relevant type*. It is easy to see that evolution could favor truth (or correctness or accuracy, if a less honorific notion is needed) and information content, because beings whose mental representations are superior in these qualities are likely to survive and leave progeny. As Richard Boyd and Hilary Putnam have argued more recently, truth is instrumental in the beings' efficiency of attaining their goals. But a similar story is more difficult to tell for aesthetic properties, precisely because it is not immediately evident that possession of a simpler theory gives the beings an advantage.[19] There is, however, a possibility. Human beings have finite mental capacities. It is arguable, then, that a computationally simple or easily applicable theory gives a finite being an advantage. But as several writers (including Schlick) have observed, this notion does not coincide with simplicity as parsimony of primitive symbols or irreducible laws—it is not the notion of *cognitive* economy we are after.[20]

The problem, then, is that the biologically inculcated simplicity rankings may prove to be useless in more theoretical matters. Schlick himself observes that man's capacity of judgment and inference enhances his chances of survival: in a sense he must rise above animal instinct to hold his ground in nature.[21] But does this not mean that the biologically grounded tastes have been replaced by theoretical ones? Quine, a fellow naturalist, conjectures that the preference for simplicity is a built-in feature of human cognitive capacity, spinning on the same conceptual orbit with our innate (and evolving) similarity measures, with our sense for kinds, and with our primitive inductions. That our innate and subjective quality spacings are in harmony with nature's own groupings is for Quine a fact. But though our primitive inductions tend to be right, a fact in part explainable by evolution and natural selection, they have their failures in more theoretical matters.[22] This explains why, in Quine's view, nature has endowed us not just with inborn similarity measures but also with reason and hence the possibility of rising above them.[23] But the way from Quine's animal reason to reason thus understood is so tortuous that it is hard to see how, say, theory choices could have evolutionary explanations.

Finally, once we replace *a priori* reason with its mundane evolutionary counterpart, we not only naturalize but also pragmatize our tastes—and lose the normative aspect so dear to Kant and Schlick. To see this, consider a recent revival of Kantian themes in the historicist proposal of Thomas Kuhn. According to him, scientists throughout history have entertained a number of desiderata (such as accuracy, scope, simplicity, and fruitfulness), but interpreted and weighted them differently.[24] These desiderata are akin to norms and values which must be balanced in specific choices, and Kuhn maintains that there must be, among objective criteria, subjective and even idiosyncratic criteria for choices to be possible at all.

Like Kant before him, Kuhn allots judgment a fundamental role in the mechanism of theory choice. And like Schlick he stresses that the absence of algorithmic rules for theory choice does not mean that there are no reasons, even good ones, for particular choices.[25] But although Kuhn insists that such judgments are not matters of taste he at the same time does think that they hinge on a communal verdict. Kuhn is a historicist, unlike Kant and Schlick: the intersubjective validity of these judgments is intraparadigmatic and not universal (or interparadigmatic). Although Schlick observed that "simplicity is a concept half pragmatic and half aesthetic,"[26] he did not relativize judgments of its application to a scientific community or individual. And although he acknowledged that there can be no compelling reason for preferring the Copernican system over its Ptolemaic rival, he did not think this was because simplicity is vague or ambiguous but because adopting the former required abandonment of deeply rooted intuitions. Similarly, Schlick contended that Einstein's theory really is simpler than that of Lorentz. Both theories can be regarded as true, but, just as all our past experience testifies to the truth of the Copernican view, so "all previous experience likewise shows that Einstein's principles (of the relativity of lengths, simultaneity, etc.) are true of our actual world."[27] The reason for the underdetermined choice was not in two rival notions of simplicity, but in the necessity to weigh simplicity against cherished (if no longer *a priori*) intuitions of absolute space and time.

Kuhn's recent replies to allegations of irrationality and relativity lean towards the following principle: given two theories which are equal in all desiderata but one, scientists do and should choose the theory which excels in this remaining desideratum.[28] This principle of rationality appears to be both descriptive and normative, and valid for all desiderata, aesthetic or not. Its differentia is that the desiderata are not grounded by a transcendental deduction or evolutionary justification: we have simply come to identify the enterprise of science as an activity in which they are sought.[29] Such a notion of rationality would have been too thin to the tastes of Kant and Schlick.

Boston University and The Academy of Finland

NOTES

1. Kant 1968, A645=B673.
2. The unity of the end to which all the parts relate and in the idea of which they all stand in relation to one another, makes it possible for us to determine from our knowledge of the other parts whether any part be missing, and to prevent any arbitrary addition, or in respect of its completeness [to discover] any indeterminateness that does not conform to the limits which are thus determined a priori. The whole is thus an organized unity (*articulatio*), and not an aggregate (*coacervatio*). It may grow from within (*per intussusceptionem*), but not by external addition (*per appositionem*). It is thus like an animal body, the growth of which is not by the addition of a new member, but by the rendering of each member, without change of proportion, stronger and more effective for its purpose. Kant, *op. cit.*, A832-3=B860-61.
3. See Rescher 1983, p. 89.
4. Reason's craving for structured unity is, like sensibility's insistence on (absolute) space and time, and understanding's reliance on the category of causation, a built-in feature of human intelligent. It is, in principle, possible that there are beings with different types of intelligences. But for us this remains an abstract possibility. Although Kant does not intend to show that all rational beings favor a body of knowledge with the required architechtonic beauty, he maintains that all creatures with our form of sensibility and our categories of understanding and faculty of reason akin to ours necessarily agree. Kant

is here expressly anthropomorphic, but normative at the same time: the whole point of the Copernican Revolution is here. For a discussion of these aspects, see Rescher, *op. cit.*, Chapter III.

5. Kant 1952, Introduction, Section III.

6. *Ibid.*, pp. 17, 179.

7. *Ibid.*, pp. 27, 187-88.

8. *Ibid.*, p. 220.

9. *Ibid.*, Introduction, p. 188.

10. Historically the most influential link between Kant and Schlick is no doubt Ernst Mach and his idea of the economy of thought. However, space does not allow me to deal with Mach.

11. Schlick 1925, p. 13. Interestingly, Schlick observes what is so difficult about explanation: it is not enough that we find a few explanatory principles. Rather, it is also required that we can subsume the variety of phenomena under these principles, in all their detail: "we are called upon to employ a minimum of explanatory principles, and at the same time to determine completely with their aid every single phenomenon in the world." And he refers to the mode of knowledge of an untutored person as "erudition" (*Wissen*), and that to be found in scientific understanding as knowledge (*Erkenntnis*). The former consists of unconnected pieces of knowledge: The mind of an untutored person "is set at rest as soon as *some name or other* is assigned to each thing or phenomenon . . . . How often do we hear people priding themselves on their storehouse of names, phrases and numbers, which they would pass off as knowledge." *Ibid.*, p. 15.

12. Schlick 1915, pp. 169-70.

13. *Ibid.*, p. 171. Another writer impressed by Kant was William Whewell who gave an essentially Kantian characterization of progress in science. Whewell called progress of unification which resulted when empirical laws were derived from a more fundamental law "Consilience of Inductions." See Whewell 1847. For a modern discussion of unification, see Friedman 1974 and Sintonen 1984.

14. Einstein wrote, on December 9, 1919, to Max Born, asking for help in attempts to get a chair to Schlick. Einstein foresaw difficulties, because, as he put it, Schlick "does not belong to the established church of the Kantians." *The Born-Einstein Letters*, New York: Walker & Co., 1971. Quoted from Feigl and Blumberg (1974), XIX.

15. Schlick 1915, p. 163.

16. Schlick 1909.

17. *Ibid.*, p. 1. Schlick here sets out to make ethics an empirical science, and the main objection he must address is the one posed by Wilhelm Wundt who had questioned the very intelligibility of this rendering of "why": "The question of why we feel pleasure, displeasure, and so on, is thus really just as vacuous as if we sought to ask why we touch, smell, taste and so forth, and why utterly different sensations have not developed instead of these." Now according to Schlick to explain a fact always "amounts to displaying the latter as a special case of another fact."

18. Schlick (1974), p. 94 and 101.

19. As, e.g., Richard Boyd 1980 and Hilary Putnam 1978 have argued more recently, the realist notion of truth might well receive an explanation through the fact that a truer picture of the world contributes to better chances of obtaining goals. Hence it makes sense

to think that a truthlike (truer and more informative account) is preferred by Mother Nature, and that truth as a cognitive value gets explained.

20. Mary Hesse labels computational ease and other pragmatic notions of simplicity subjective. For discussion, see Hesse 1974, Chapter 10.

21. Schlick 1925, p. 95.

22. "Creatures inveterately wrong in their inductions have a pathetic but praiseworthy tendency to die before reproducing their kind." Quine 1969, p. 13.

23. What works in one area is a hindrance in another, writes Quine. Therefore "the immediate, subjective, animal sense of similarity" gives way to "the remoter objectivity of similarity determined by scientific hypotheses and posits and constructs." In this theoretical sense, Quine says, things are similar to the extent they are "interchangeable parts of the cosmic machine" uncovered by evolving science. Quine, *ibid.*, p. 19.

24. For discussion, see especially Kuhn 1977.

25. Schlick quotes approvingly (though he adds a qualification) Cassirer who had criticized the view that equivalent theories have equal logical justification: "The defect of this inference, however, is plain; for the abolition of an absolute standard in no way involves the abolition of differences in value between the various theories." Schlick 1915, p. 172.

26. Schlick 1931, p. 182.

27. Schlick 1915, p. 171.

28. See, e.g., Kuhn 1983, p. 564.

29. My suspicion is that, if forced to choose, Kuhn would lean towards a naturalistic explanation of some of these virtues, because exemplars are crucial for scientific training and later research, and because exemplars embody similarity measures and schemata of classification. The idea that these measures are results of evolution is then a natural one. But my suspicion is also that for Kuhn the evolution (and revolution) makes strictly biological similarity measures obsolete.

REFERENCES

Boyd, Richard. 1980. "Scientific Realism and Naturalistic Epistemology." In P. D. Asquith and R. N. Giere, ed., *PSA 1980*, Vol. 2. East Lansing, Michigan: The Philosophy of Science Association.

Feigl, H. and Blumberg, A. E. 1974. "Introduction," in Moritz Schlick, *General Theory of Knowledge*, New York and Wien: Springer Verlag.

Friedman, Michael. 1974. "Explanation and Scientific Understanding." *The Journal of Philosophy*, vol. 71, pp. 05-19.

Hesse, Mary. 1974. *The Structure of Scientific Inference*. London and Basingstoke: The Macmillan Press.

Kant, Immanuel. 1952. *The Critique of Judgment*. Oxford: Clarendon Press.

Kant, Immanuel. 1968. *Critique of Pure Reason*. Tr. by Norman Kemp Smith. New York: St. Martin's Press.

Kuhn, Thomas. 1977. "Objectivity, Value Judgment, and Theory Choice." In T. Kuhn, *The Essential Tension*. Chicago: University of Chicago Press, pp. 320-39.

Kuhn, Thomas. 1983. "Rationality and Theory Choice." *The Journal of Philosophy*, vol.

80, pp. 563-70.

Putnam, Hilary. 1978. *Meaning and the Moral Sciences*. Boston, London and Henley: Routledge & Kegan Paul.

Quine, Willard Van Orman. 1969. "Natural Kinds." In Nicholas Rescher (ed.), *Essays in Honor of Carl G. Hempel*. Dordrecht: D. Reidel, pp. 01-23.

Rescher, Nicholas. 1983. *Kant's Theory of Knowledge and Reality, A Group of Essays*, Washington, D.C.: University Press of America.

Schlick, Moritz. 1909. "The Fundamental Problem of Aesthetics Seen in an Evolutionary Light." In Moritz Schlick, *Philosophical Papers, Vol. I (1909-1922)*. Ed. by H. L. Mulder and B. F. B. van de Velde-Schlick. Dordrecht and Boston: D. Reidel, 1979.

Schlick, Moritz. 1974. *General Theory of Knowledge*. Second edition. Tr. by Albert E. Blumberg. Wien and New York: Springer Verlag. Originally published in 1925 as *Allgemeine Erkenntnislehre*.

Schlick, Moritz. 1915. "The Philosophical Significance of the Principle of Relativity." In Moritz Schlick, *Philosophical Papers, Vol. I (1909-1922)*. Ed. by H. L. Mulder and B. F. B. van de Velde-Schlick. Dordrecht and Boston: D. Reidel, 1979.

Schlick, Moritz. 1931. "Causality in Contemporary Physics." In Moritz Schlick, *Philosophical Papers, Vol. II (1925-1936)*. Ed. by H. L. Mulder and B. F. B. van de Velde-Schlick. Dordrecht and Boston: D. Reidel, 1979.

Schlick, Moritz. 1962. *Problems of Ethics*, New York: Dover Publications.

Sintonen, Matti. 1984. *The Pragmatics of Scientific Explanation*, Acta Philosophica Fennica, Vol. 37, Helsinki 1984.

Whewell, William. *The Philosophy of the Inductive Sciences, Founded upon Their History*, Second Edition, London: Frank Cass & Co., 1847, Book II.

SIMPLICITY AND DISTINCTNESS

Ulrich Majer

BEFORE I jump *in medias res* and explain Hertz's notion of "simplicity and distinctness" as an instance of aesthetic reasoning in science, let me make some brief introductory remarks: (1) It seems to me—this conference notwithstanding—that most philosophers of science, in particular those in the logical empiricist tradition, still tend to ignore aesthetic factors in science as a matter of mere pragmatic relevance which has no impact on the development of science in the long run. (2) This empiricist tendency stands in an obvious and strong contrast to what scientists themselves have to say about the role of aesthetic considerations in science. The following quotations are chosen rather at random and would be even more impressive if I had included more remarks from scientists around the turn of the last century, when "simplicity" was in fashion as a main topic:

> In der Naturwissenschaft versuchen wir das Spezielle aus dem Allgemeinen abzuleiten; das Einzelphänomen soll als Folge *einfacher* allgemeiner Gesetze verstanden werden. Die allgemeinen Gesetze können, wenn sie sprachlich formuliert werden, nur einige *wenige* Begriffe enthalten, denn sonst wäre das Gesetz nicht *einfach* und allgemein. (Werner Heisenberg. "Sprache und Wirklichkeit" in *Physik und Philosophie* (Stuttgart: Hirzel, 1959).)

> [Die geistige Entwicklung der Naturwissenschaften, ihre Grundtendenz] selbst ist gar nichts weiter als die zunehmende *Vereinfachung und Vereinheitlichung* unseres Gedankenbildes vom Ablauf der Vorgänge in der Welt. (Erwin Schrödinger: *Vom Geist der Naturwissenschaft.*)

> Just as i=V-1 was introduced to preserve in *simplest* form the laws of algebra . . . , just as ideal factors were introduced to preserve the *simple laws* of divisibility for algebraic whole numbers . . . ; similarly to preserve the *simple rules* of ordinary Aristotelian logic, we must supplement the finitary statements with *ideal elements.* (David Hilbert: "On the Infinite," in *Philosophy of Mathematics* (Englewood Cliffs: Prentice Hall, 1964), p. 195.)

> In a law of nature, as we shall later establish more precisely, *simplicity is essential.* The assertion that nature is governed by strict laws is devoid of all content if we do not add the statement that it is governed by mathematically *simple laws.* This matter is somewhat analogous to the fundamental law of multiple proportions in chemistry: it loses all its content unless we add that the combination occurs in small integral multiples of the relative atomic weights. That the notion of law becomes empty when an arbitrary complication is permitted was already pointed out by Leibniz in his *Metaphysical Treatise.* Thus *simplicity* becomes a working principle in

the natural sciences. (Herman Weyl: *The Open World* (New Haven: Yale University Press, 1932), p. 41.)

Far from being superficial remarks concerning merely pragmatic aspects of science the remarks instead reveal that scientists take simplicity and similar aesthetic notions to be of deepest concern for an adequate understanding of science; this is in particular true of the last two remarks because Hilbert and Weyl as mathematicians did know that without some restrictions the possibilities of the mathematical description of the world would be indefinite.

Now, instead of criticizing the logical empiricist view I will rather explain why aesthetic considerations play an essential role in science, that is, in our conscious thinking about nature. The thesis which I want to defend is the thesis that the only notion of truth which is empirically available to us, namely the notion of correspondence with the phenomena, is insufficient to determine the choice among different yet empirically equivalent theories; therefore, there is the need for non-empirical considerations, that is, in a broad sense, the need for aesthetic considerations in science.

The need for aesthetic considerations in science was to my knowledge first stressed by Heinrich Hertz, the discoverer of radio waves and the electro-magnetic nature of light, and I will now explain, according to his concept of scientific reasoning, why and how aesthetic criteria play an indispensable role in our choice among different, yet empirically equivalent, theories. It is obvious that the answer to this question depends strictly on the answer to another question, namely the question "What do we understand by a theory?", and for this reason I first have to explain Hertz's conception of theories. Because this conception evolved from a critique of Kant's philosophy of science let me, before I start the systematic part, make some remarks about Kant and the historical situation in the second half of the nineteenth century.

I. The Transformation of Kant's Philosophy in the Nineteenth Century

In the second half of the nineteenth century, when the reign of Fichte, Hegel, and Schelling declined, a discussion took place among philosophically minded scientists in the course of which Kant's philosophy became not so much rejected as *transformed.* To this context belong, among others, the works of Hermann von Helmholtz and Heinrich Hertz, who both tried to reconcile Kant's transcendental philosophy with the methods and results of modern science, which appeared in conflict with Kant's theory of mathematics and pure physics as a corpus of *a priori synthetic judgments.* The basic strategy was roughly this: Instead of rejecting Kant's doctrine of *a priori* synthetic judgements directly—as most mathematicians and physicists of the time did[1]—they criticized the underlying assumption of the two sources of all our knowledge: intuition and thinking, which according to Kant are both *necessary*—"thoughts without intuitions are empty, intuitions without concepts are blind"[2]—and at the same time mutually *exclusive*, insofar as concepts can only be thought and not intuited and objects can only be intuited and not be thought.[3]

The way in which Helmholtz and Hertz criticized the underlying distinction of the two sources of our knowledge is not so much a direct attack on one of its characterizations, the mutual exclusion of intuition and thinking, but by launching an alternative program of epistemology, which made no use of the distinction at all, but instead tried to show that all our knowledge and beliefs about the external world rests on one and the same *process*, namely that of "symbolization," which cannot be separated into two distinctive parts.[4] This theory of "symbolization," as I will call it, was developed in two steps, the first taken by Helmholtz, the second by Hertz; let me explain this briefly:

(1) Helmholtz as a sense physiologist had the idea that the sequence of our (inner) sensations may be analyzed as a kind of natural language in which the external world of phenomena speaks to us like a sequence of imprints in a script. This linguistic program was based on two principal assumptions: (i) Sensations are not pictures but "*symbols*" of the objects (and properties) they represent. By this Helmholtz means that the qualities of our sensations like sound, heat, color, and taste are not identical or similar to the qualities of the objects they represent—as we would expect from a good picture, say, of a flower—but instead that they are mere *functions* of the sense-organ stimulated by the object; as Helmholtz put the the point somewhat paradoxically: "Light first becomes light, if it strikes a seeing eye, without this it is only an oscillation of the ether."[5]

(ii) Sensations are related to the objects of the external world by the law of causation, which means roughly that under the same circumstances in the external world always the same sensations will occur (provided the observer is in the same relation to the surrounding world). The law of causality with respect to sensations is according to Helmholtz a *transcendental* law in Kant's sense, that is, it is not the result of any experience but instead a necessary presupposition for the very possibility of experience; this will become clear in the following conclusion, which states the core of the linguistic program.

(iii) Although the sensations are not pictures but symbols of the objects they represent, the *sequences of sensations nonetheless are pictures* in the literal sense that their order is identical with (or at least similar to) the order of causes, that is, the order of phenomena and their relations by which they are brought about. In modern terminology we would characterize the relation between the order of inner sensations and their external causes as a structural isomorphism between both sequences.

(2) In passing over to Hertz I would like to stress that the term "object of the external world" has been taken in a very broad sense including not only objects in the usual commonsense (like tables and chairs) but

also properties and relations, and in general every external phenomenon that can effect my sensations. Now, this causal theory of symbolization was modified by Hertz in two fundamental respects:

(i) Hertz dropped the law of causation as a transcendental assumption for the possibility of experience and replaced it by certain *laws of coordination* between the results of observations or measurements and their symbolic representation—"law" taken here in the regulative or prescriptive sense that we should follow certain rules in co-ordinating symbols with phenomena.

(ii) The second modification is related to the first. In order to make the coordination a workable substitute for the causal relation which had warranted the picture-relation between the sequence of inner sensations and their external causes, Hertz had to modify both sides of the relation: (a) Besides sensations as a kind of natural symbols he had to admit the construction of artificial, in particular, of functional symbols like "mass" and "charge," or "force" and "field." (b) For the sake of physics as a quantitative science he had to restrict the rather vague notion of external object or phenomena to that of measurable quantities like "distance," "duration," and so forth.

It should be noted that these modifications lead to a theory of "picture-formation," as I will call it in distinction to the former theory of symbolization, in which intuition and thinking are no longer separable, because a picture in Hertz's sense entails representative as well as inferential moments completely intertwined: the pictures are not static but "*dynamical models*" of the sequences of events in the world, to use a characteristic phrase from Hertz:

> We form ourselves *images* or *symbols* of external objects; and the form which we give them is such that the *necessary consequents of the images* in thought are always the *images of the necessary consequents* of the things pictured. In order that this requirement may be satisfied, there must be a certain conformity between nature and our thought. When from our accumulated previous experience we have once succeeded in deducing images of the desired nature we can then in a short time develop by means of them, as by means of *models*, the consequences which in the external world only arise in a comparatively long time.[6]

Now, this notion of a picture as a "dynamical model," in which inference and representation are interlocked as function and argument in Frege's system of logic, is such a general one that Hertz had to restrict the set of "admissible" pictures by certain *selection rules* in order to make actual use of them. It is these selection rules which we have to study next, because they provide an answer to our main question, why and how aesthetic considerations play an essential role in science.

II. Need for Aesthetic Criteria in Science

Hertz constitutes the selection rules by formulating three "criteria" which every picture has to pass like a set of subsequent filters in order to become scientifically accepted by us. Here are the three criteria:

(a) logical consistency, that is, no contradictions against the laws of thinking
(b) empirical adequacy or correspondence with the phenomena
(c) simplicity and distinctness

Although the first criterion of logical consistency stands in a close and interesting relation to the third[7] we have to deal here only with the third criterion in its relation to the second because it is the peculiar proportion of these two which entails the answer to our question. Let me first make a brief remark about the notion of "correspondence" before I explain in which sense "simplicity" and "distinctness" are aesthetic notions.

Usually a correspondence (in the truth-theoretical sense) is understood as a relation between a picture, let's say of the Cathedral of Learning, and what is depicted, the Cathedral itself. Now, philosophers of all times, from Plato to Frege, have found severe difficulties in this notion of "correspondence" as an explication of the notion of truth: Is it an identity? This is impossible because in this case the picture and the thing depicted should be one and the same—and that is exactly what one does not want to maintain. Or is it a similarity? But then the question arises "Similar in which respect?", and here again we encounter an impasse: Are both, the picture and the thing depicted, identical or only similar in the given respect? The first answer seems impossible for the same reasons as already stated and the latter leads to an infinite regress. Or is correspondence no explication of truth at all but only a disguised intentional ingredient of our epistemic goal to recognize the truth, which has no real connection with the notion of truth, as Plato and Frege suggested?[8]

Despite these difficulties scientists never seemed prepared to dispense with correspondence as an essential part of the notion of truth, and Hertz is no exception in this regard. It is, however, very important to note that Hertz uses the notion of correspondence in a rather different way than usual:

(i) First, correspondence is only a necessary and not a sufficient condition for truth; the judgement that the correspondence is indeed "satisfied" has to be added before we are entitled to call a picture "true"; the incorporation of judgement into the complete notion of truth grants indeed Frege's strongest objection against a definition of truth by correspondence or similar concepts.[9]

(ii) Second, and more important, correspondence is not a relation between a picture and the things depicted, as they exist in themselves, so to speak, but only between the picture and the things as they *appear* from different points of view; that is to say between the phenomena and their natural relations on the one hand and the symbols and their functional connection on the other.

The second point is not only evident for pictures in the ordinary sense of the word because they show their objects always in a certain perspective but also for pictures in the scientific sense because they represent their objects always in the form of particular measurements. It would be highly interesting to show how this notion of correspondence (as isomorphism between the natural sequence of phenomena and the functional sequence of symbols) escapes the objections raised by Frege and other philosophers. However, for lack of time, I have to confine myself to presenting an example from which one can see how the notion of correspondence with the phenomena functions with respect to a scientific theory.

In the last century, as you know, several proposals have been made to deal with electro-magnetic phenomena; let me mention only the four most important: the theories of Faraday and Maxwell, Helmholtz, Hertz, and Gauss and Weber. Now, with respect to these theories the notion of correspondence with the phenomena functions as a cut: it separates the theories into two (or more) *equivalence classes*: first, the class of Maxwellian theories to which all and only those theories belong which have the *same set of phenomena as consequences*, regardless how different the theories in other respects may be; second, the (complementary) class of non-Maxwellian theories to which all and only those theories belong which have a different set of phenomena as consequences. To the first class belong, beside Maxwell's own equations, the theories of Helmholtz and of Hertz, whereas the theory of Gauss and Weber belongs to the second class.

Now, it is important to note that the theories in the first class—although they cannot be distinguished by empirical means—can be distinguished by other, *non-empirical* criteria, like those of simplicity and distinctness. And this is exactly what Hertz does when he argues that his own picture of electro-magnetic phenomena is simpler (and perhaps even more distinct) than the two other pictures in the same class, because his picture is based on the concept of "polarization" as the only fundamental notion, whereas the other two pictures use several, independent notions.[10]

Before I go on to explain in which sense simplicity and distinctness are aesthetic notions let me make one remark to prevent later misunderstandings: I agree with the logical empiricists that there are many differences between the theories of the same equivalence class which are completely trivial and irrelevant with respect to the empirical content of these theories: differences of notation, of language, of formulation, of representation, and so forth. But I cannot accept their view that all the non-empirical differences are irrelevant with respect to the choice among different theories. Some are not! And I will now explain why. If we call the former differences "differences of *style*" the view I want to defend in accordance with Hertz can be formulated as follows: The differences of "style" do not exhaust the differences among empirically equivalent theories; beside differences of "style" there are other differences too, like simplicity and distinctness, which are relevant for the choice among different (yet empirically equivalent) theories. Let us call these non-stylistic but somehow relevant differences "aesthetic" differences and ask: What is their epistemological status? What does Hertz understand by "simplicity and distinctness" as instances of aesthetic differences among theories?[11] The answer Hertz gives is rather short, perhaps too short to be intelligible; nonetheless let me first quote the answer and then elaborate its bearing on our main topic:

III. The Role of Aesthetic Factors in Science

Having explained the selective role of the first two criteria Hertz continues:

> [But] two permissible and correct images of the same external objects may yet differ in respect of *appropriateness*. Of two images of the same object that is the more appropriate which pictures more of the essential relations of the object—the one which we may call the more *distinct*. Of two images of equal distinctness the more appropriate is the one which contains, in addition to the essential characteristics, the smaller number of superfluous or empty relations—the *simpler* of the two. [And Hertz adds the remark that] empty relations cannot be altogether avoided.[12]

Having got a rough idea what "simple" and "distinct" mean the first point one must understand (which isn't very well expressed in the English translation) is the circumstance that both criteria are not completely independent of each other: improvement of distinctness may diminish the simplicity of a picture and vice versa; for example, the introduction of "epicycles" into the Ptolemaic system enables us to explain the retrograde movements of the inner planets and thus the picture becomes more distinct, but at the same time also less simple because we do not believe epicycles to be real, that is, we do not believe that they represent real movements of the planets.

The mutual interdependence of simplicity and distinctness poses an *optimalization problem*, which in general has no unique solution: one picture may be simpler or more distinct than another and nevertheless both pictures can have the same degree of appropriateness due to a compensation in simplicity or distinctness. Thus for example Hertz's own picture of mechanics is simpler than Newton's because it does not use the concept of *force*, but for the same reason it is less distinct because Hertz had to substitute the intuitive concept of force (forces can be graphically represented by vectors) by very complex boundary conditions, called "holonomous systems," which are not intuitive at all; therefore, both pictures have roughly the same appropriateness.[13]

The *multiplicity* of possible solutions in respect to simplicity and distinctness leads thus to an underdetermination in the choice among different yet empirically equivalent theories with roughly the same appropriateness, which in turn introduces the need for *aesthetic judgements* in science. This is most clearly expressed by Hertz himself:

> We cannot decide without ambiguity whether an image is appropriate or not; as to this *differences of opinion* may arise. One image may be more suitable in one direction, another in another; only by gradually testing many images can we finally succeed in obtaining *the most appropriate*.[14]

The next point is more difficult to grasp because it is related to the semantic question: What does Hertz understand by "essential relations of external objects," in respect to which one picture is simpler or more distinct than another? To put the same question differently: How is it possible that two empirically equivalent pictures can still be different in simplicity or distinctness or both? Are the "essential relations," and with them their symbolic representation, not uniquely determined by the requirement of correspondence? The answer is rooted, I think, in Hertz's conviction that truth in the empirically scrutable sense of the word is restricted to correspondence with the phenomena and that we cannot transcend this limit without leaving the domain of knowledge and entering that of belief:

> The images which we here speak of are our ideas of things. With the things themselves they are in agreement in *one* important respect, namely, in satisfying the above-mentioned requirement [of correspondence]. Yet it is not necessary for fulfilling their purpose that they should be in agreement with the things in any other respect whatever. As a matter of fact, we do not know, nor have we any means of knowing, whether our ideas of the things are in correspondence with them in any other than this *one* fundamental relation.[15]

To unpack the quotation I have to make a distinction between "external objects" on the one hand and "things (in themselves)" on the other, which is not explicit in the quotation above but justified by Hertz's own procedure: (i) "external objects" in the logical sense of coordination of symbols to something "given" are the *phenomena* themselves, which, according to Hertz, are the results of certain measurements specified in the coordination rules;[16] (ii) "things" in the ontological sense of "existence" are the time and space independent "bearers" of the phenomena; they are constructed by us by means of formulating "laws" among the phenomena, and we say "a thing exists" just in case we believe the laws hold for all past and future instances; it should be noted that existence-sentences like "there exist electrons" is not the expression of a judgement but of a belief! "There exist electrons" means we believe that the laws of electrodynamics hold (together with some other laws).

An analogous distinction can be made between "external relations" on the one hand and "constructed relations" on the other: (i′) "external relations" in the mathematical sense of a structural isomorphism between the sequence of (inner) symbols and the sequence of (external) phenomena are the time dependent functions among the phenomena themselves, "function" taken here in the strictly extensional sense of a progression from one phenomenon to its associated next; (ii′) "constructed relations" in the ontological sense of existence are the properties and relations among the things constructed; their meaning, although deductively connected with the phenomena, is not purely extensional but in part intensional and theory dependent. To say, for example, that gravity exists is to express the belief that certain bodies will move in certain ways and thereby give rise to such and such phenomena. One should remark in this context that the existence of gravity as a relation among "constructed" things implies the existence of "heavy bodies" and *vice versa*!

With these distinctions in mind we can now see why empirically equivalent pictures can still differ in simplicity and distinctness. To take the latter first: If we understand by "essential" relations those (external) relations among the phenomena, which are *invariant* under space-time and other transformations, a picture can fail to make them explicit (though implicitly it entails these relations); take for example Gauss's *Principle of Least Constraint*: From this one cannot see immediately, as in Newton's picture, that acceleration is the principal constant in all phenomena of motion, and therefore this picture may be less distinct than Newton's. On the other hand, it has been doubted, at least since D'Alambert and Lagrange, that Newton's "forces at a distance" are real, and for reasons which are primarily connected with the rise of electrodynamics as a field theory, Kirchhoff and Hertz tried to show that "force" is an empty term and to substitute for Newton's picture a simpler one: "In half a page"—as Boltzmann expresses his disagreement—"forces had been defined away and banished from nature and physics made into a descriptive science properly speaking."[17]

The possibility of drawing different pictures within the same class of empirically equivalent theories is, however, by no means restricted to mechanics; take for example thermodynamics as a phenomenological theory of heat and related phenomena: Here it can even be proved by Poisson's theory of total differential equations that beside "entropy" different other functions exist, like "enthalpy" and the "Planck-Massieu-function" and so on, in which thermodynamics not only can be but indeed has been formulated. Furthermore these different formulations differ in appropriateness for different applications: for example, a refrigerator is easier to describe in terms of enthalpy than in terms of entropy.

IV. Between Instrumentalism and Realism: Aesthetic Reasoning in Science

Having shown that different yet empirically equivalent pictures not only can but in fact do exist (for most of the known theories) I will now return to the question: Why should we welcome such a pluralistic situation instead of trying to overcome it? There are at least two principal objections or alternatives to our pluralistic position, one on the extreme left, the other on the right, so to speak: radical empiricism (or instrumentalism) and semantic realism.

(1) Radical empiricism ignores all the differences between empirically equivalent pictures as mere linguistic, pragmatic, or even worse, as sheer metaphysical rubbish, and maintains that theories are nothing else but instruments for the deduction of possible observations, such that two theories are different if and only if they have different sets of observable consequences. This hermetic position is to my mind irrefutable—although it seems more a caricature of the real situation than a true description. Here, I will not deal with it because it has been criticized so often by Popper, Feyerabend, and Kuhn[18] that I could hardly offer any new argument, except perhaps the following two remarks:

(a) For radical empiricism the Craig-sentence C_T of a theory should be as acceptable as the theory itself from which it has been constructed because the Craig-sentence belongs to the same equivalence class as the original theory T. This, however, is *absurd* because the Craig-sentence produces the observational consequences by a physically absolutely crazy algorithm: Gödel-numbering.[19]

(b) If theories have an irreducible hypothetical status (as I have argued elsewhere[20]) the problem of induction becomes unsolvable in the following probabilistic sense: The estimation of the relative probability (of truth) of a theory, given any finite amount of evidence, presupposes that the truth of the theory as a whole is a *function* of its arguments, the singular observation sentences. This, however, is not the case, if the theory is constructed in the logical form of a general hypothesis including an existential assumption, because the terms "for all" and "there exists" denote no real truth function from which one could estimate the relative probability of the theory, but indicate the presence of a conviction or a belief. This belief and not a fictive order of objective degrees of relative probabilities determines our choice among different theories.

(2) The second alternative to our pluralistic position is semantic realism,[21] the best representative of which, I think, is still Frege. Therefore, let us see what he has to say to our pluralistic position.[22]

(a) The first and perhaps most astonishing aspect is that Frege addresses the pluralistic position at all and furthermore seems to confirm our interpretation of Hertz as a "phenomenalist"—at least indirectly—insofar as Frege discusses the phenomenalism of a group of scientists to which Hertz doubtless belongs.

In his late essay "The Thought" (published in 1918, after Frege had read an early version of Wittgenstein's *Tractatus*) one finds a paragraph beginning: "It is strange how, in the course of such reflections, opposites turn upside-down."

In this and the subsequent paragraphs Frege attacks a group of scientists, to which he refers anonymously by the circumscription "a sense-physiologist," to which not only and obviously Mach and Helmholtz, but also Hertz belongs. This attack is of particular interest for us, because Frege wants to prove by a kind of *reductio ad absurdum* of the epistemological position of the "sense-physiologist" that we are able to "judge." And "judgement" means for Frege the transition from sense to reference, from a thought to its associated truth-value; the latter implies in particular the transition from a sense of a name to the designated object. Hence, the possibility of judging depends, at least for us human beings, on our ability to *recognize* objects. Can we recognize objects—in particular objects in the external world? Here, the sense-physiologist steps in and says: No!—at least not in the sense of a categorial knowledge but only, if at all, in the sense of a hypothetical belief, because all our knowledge is restricted to the realm of phenomena, to the sphere of the senses, as Frege calls it.[23]

(b) Now, it is very important to recognize that Frege proves the possibility of judgment *indirectly* by proving that the epistemological assumption which the sense-physiologist makes, namely the assumption "that only what is my idea can be the object of my awareness," leads to the "quite incredible" consequence of *solipsism*: that "all my knowledge and perception is limited to the range of my ideas, the stage of my consciousness. In this case I should have only an inner world and I should know nothing of other people." Because this "inevitable consequence" is absolutely unacceptable (for Frege), the premiss must be false "that only what is my idea can be the object of my awareness." From this indirect proof Frege jumps (after some intermediate steps to which I will come back in a moment) to the final conclusion that we not only "*can* venture to judge about things in the external world"—but furthermore—that "we *must* make this venture even at the risk of error if we do not want to fall in far greater dangers"—namely that of solipsism.

(c) Is Frege's proof conclusive? Do we really recognize "things in the external world" in the sense that we have a knowledge of them that goes beyond their appearance to the things themselves, so to speak, as Frege supposes with his explanation of "judgement" as a transition from the sphere of sense to that of reference? I do not think so and I will now explain why:

From mathematics we know that on many occasions we introduce "*ideal elements*" into a given structure whenever the need arises to keep the theory of that structure "simple and distinct." (Remember the corresponding remark by Hilbert, quoted at the beginning.) To mention only a few well known examples: In arithmetic we introduce "negative" and "irrational" numbers, in geometry "points" and "lines," in analysis "cuts" and "continuous fields," and so on. But at the same time we know that these elements have only an *ideal existence*, like "unicorns" or "thunderbolts" in a story, and that we have to be careful to the utmost in judging whether these things *really exist* or not.

So, for example, it is a big question whether points and lines really exist outside geometry in the independent world of perceptible things, or whether they have only a contingent existence as ideal constructions within geometry. A similar question arises in regard to negative and irrational numbers: Are they only the product of certain operations of our mind, such that it only makes sense to speak of them within the realm of these operations, or have they a comparatively independent existence like those of natural numbers, where the adjective "natural" indicates their theory-independent status? I will not answer these questions now, but only direct the reader's awareness to the fundamental difficulties involved.

(d) Frege himself, I think, was fully aware of these difficulties because he discusses in the intermediate step, which leads to the final conclusion, a difficulty in connection with our judgements about the external world, which intimately resembles the above mentioned question of real existence:

> But may this [the recognition of something which is not an idea] not be based on a mistake? I am convinced that the idea I associate with the words "my brother" corresponds to something that is not my idea and about which I can say something. But may I not be making a mistake about this? *Such mistakes do happen. We then, against our will, lapse into fiction. Yes, indeed!* By the step with which I win an environment for myself I expose myself to the risk of error. And here I come up against a further difference between my inner world and the external world. I have no doubt that I have a visual impression of green, but it is not certain that I see a lime-leaf. So, contrary to widespread views, we find certainty in the inner world, while *doubt never altogether leaves us in our excursions into the external world.*[24]

Thus, Frege himself is sceptical in regard to our knowledge about the external world, and it is exactly this intermediate position of *scepticism* which he left out when he jumped from the disproof of solipsism to its immediate opposite: our ability to judge, that is, to make the transition from the sphere of sense to that of reference. But between both extremities, solipsism and realism, is the intermediate position of scepticism with respect to our knowledge about the external world, and what Frege really proved is to my mind a much weaker claim, namely the claim that in order to avoid solipsism we have to recognize a "third realm," the realm of thoughts, which neither belongs to our subjective ideas nor to the things in the external world as they exist in themselves, but which is nonetheless objective in the sense that we have no influence upon it.

But this is something Hertz also admits insofar as that to which a picture corresponds is something extremely close to a thought: the time and space invariant relational structure between the phenomena.[25] Therefore, the principal difference that remains is in regard to the question of judgement: Can we go any step further than the acknowledgement that our pictures correspond to the phenomena? And in particular can we transcend the sphere of sense and reach that of reference? The answer, I think, is "no" insofar as knowledge is concerned, but "yes" if *hypothetical reasoning* is taken into account, which means that with respect to the *existence of things* (and theoretical entities in general) we are leaving the domain of knowledge and entering that of belief. Here, our steersman is not so much truth as primarily aesthetic values like simplicity and distinctness.

University of Göttingen

NOTES

1. Take for example Dedekind's and Frege's critique of Kant, who both tried to show that mathematical judgements are *analytic* in the sense that arithmetic, including analysis, could be based on pure thinking without any recourse to geometrical intuitions, as an extension of pure logic, so to speak, in the broad sense of the term including set theoretical notions and principles. The critique of Kant's theory of space and time and hence of pure physics as a discipline of synthetic judgements *a priori* is too well known to be repeated here.

2. Kant, *Critique of Pure Reason*, A51/B76.

3. This is not quite true because, according to Kant, objects too can be thought, yet only indirectly in relation to the immediately given intuition; see *Critique* (A50, 51/B74, 75), where Kant says: "*Ohne Sinnlichkeit würde uns kein Gegenstand gegeben, und ohne Verstand keiner gedacht werden,*" and few lines later he demarcates both faculties more precisely: "*Der Verstand vermag nichts anzuschauen, und die Sinne nichts zu denken.*"

4. At this point I want to mention my indebtedness to Paul Hoyningen, who convinced me that the distinction as such could not be criticized but only its characterizations; after all, the non-separability of intuition and thinking is not the rejection of the distinction but, if correct, the confirmation of Kant's suggestion that both, intuition and thinking, "arise from a common, but to us unknown root"—perhaps the process of symbolization?

5. Hermann von Helmholtz, "Über das Sehen des Menschen" (1855) in *Vorträge und Reden*, Vol. I, (Braunschweig: Vieweg & Sohn, 1903), p. 98; the translation is mine.

6. Heinrich Hertz: *The Principles of Mechanics* (New York: Dover, 1956), p. 1, the book contains also a chapter on "Dynamical Models," in which Hertz explicitly remarks that "the relation of a dynamical model to the system of which it is regarded as the model, is precisely the same as the relation of the images which our mind forms of things to the things themselves" (p. 177).

7. I can only indicate the relation; it is roughly the following: in a picture (theory) we combine different attributes in a single symbol such that the corresponding concept becomes more or less simple and distinct. This accumulation of attributes can deteriorate in the sense that some of the attributes contradict each other such that the corresponding concept becomes inconsistent. As a case in point Hertz discusses at length the concept of "force" of which he says:

> We have accumulated around the terms "force," and "electricity" more relations than can be completely reconciled amongst themselves. We have an obscure feeling of this and want to have things cleared up. Our confused wish finds expression in the confused question as to the nature of force and electricity. But the answer

which we want is not really an answer to this question. It is not by finding out more and fresh relations and connections that it can be answered; but by removing the contradictions existing between those already known, and thus perhaps by reducing their number. (Hertz, *Principles of Mechanics*, *op. cit.*, pp. 7-8.)

8. This is a very brief sketch of the objections given by Frege in his essay "The Thought" against the notion of correspondence as a definition of truth; although I agree with Frege that his arguments rule out certain types of "naive" correspondence theories as hopelessly muddled I do not think that his arguments are successful against a more sophisticated version of correspondence theory like the theory advocated by Hertz. See Gottlob Frege, *Logische Untersuchungen*, ed. by Günther Patzig (Göttingen: Vandenhoeck & Ruprecht, 1966).

9. A careful examination of Frege's arguments reveals that his main objection against the correspondence theory of truth is the objection that truth cannot be defined at all because the term "true" is not the name of a concept but the expression of a judgement.

10. This is a slight oversimplification insofar as other aspects than the mere number of basic notions play a role too; e.g., Hertz's picture is the only picture which is free from any "action at a distance" and therefore simpler too because it avoids a dubious notion.

11. There may be other aesthetic differences beside "simplicity and distinctness" as for example beauty or elegance, perfection, uniformity, compactness, completeness and so on and so forth. Insofar as these are genuine aesthetic and not logical notions they should be considered too; the task of this essay, however, is not the analysis of aesthetic notions in use but to make a systematic proposal how to use two of these notions.

12. H. Hertz, *The Principles of Mechanics*, *op. cit.*, p. 2.

13. That Hertz eliminated the concept of "force" from mechanics not without an appropriate substitute to compensate the otherwise much greater loss in distinctness shows that Hertz is not an eliminationist in the positivistic sense of the term: one who wants to eliminate any theoretical term from science, regardless what price he has to pay in distinctness and other aesthetic features, as in the case of the so-called Craig-elimination.

14. H. Hertz, *The Principles of Mechanics*, *op. cit.*, p. 3; it is worthwhile to remark that the German grammar reveals an unambiguous plural at the end of the quoted text.

15. H. Hertz, *The Principles of Mechanics*, *op. cit.*, pp. 1-2; this is my own translation.

16. We make—according to Hertz—our inner ideas time, space, and mass into "symbols for objects of external experience" by specifying certain "coordination-rules" how to measure these quantities in an objective, that is, in an observer-independent way by using some natural processes and bodies as measuring-standards and fix-points. It is evident from these rules that Hertz advocates a kind of objective (in distinction to a subjective) phenomenalism; see *The Principles of Mechanics*, *op. cit.*, Book II, p. 140.

17. Ludwig Boltzmann "On the Method of Theoretical Physics" (1882) in *Theoretical Physics and Philosophical Problems*, ed. by Brian McGuiness (Dordrecht and Boston: Reidel, 1974).

18. See for an excellent review Frederick Suppe's anthology *The Structure of Scientific Theories* (Urbana: University of Illinois Press, 1974).

19. For a more detailed discussion of this argument see my paper "The Craig-Elimination: A Proof for the Indispensability of Theoretical Terms" in *Proceedings of the Seventh Congress for Logic, Methodology, and Philosophy of Science* (Salzburg, 1983). The point

is, expressed in terms of simplicity and distinctness, that the Craig-sentence has lost all the distinctness of the original theory by substituting for the physical proof-theoretical relations.

20. The subsequent argument is developed in its intuitionistic core in my essay: "Ramsey's Conception of Theories: An Intuitionistic Approach," which will appear in *History of Philosophy Quarterly*, vol. 6 (1989).

21. The term "realism" was redefined by Michael Dummett to characterize a semantical position similar to Frege's, which assumes the "principle of bivalence": that every thought has one of two truth-values, independent of our knowledge, even in the case that it is not verifiable in principle. Dummett has been criticized, among others by Hans Sluga, for his characterization of Frege's philosophy as realism (in distinction to idealism). Although there is a grain of truth in Sluga's critique, I think, that Dummett's characterization of Frege's semantics is in its main line correct, because Frege defends the principle of bivalence and with it the ontology of sense and reference against any sceptical critique.

22. The following discussion is primarily based on Frege's late essay "The Thought," *op. cit.*, from which I took all the subsequent quotations. The justification for this otherwise strange procedure lies in the remarkable circumstance that Frege in this essay for the first time seems to acknowledge that his former arguments, that the sense of an indicative sentence must be the thought expressed, had a severe gap: they did not rule out the possibility that the sense of a sentence could be the ideas associated with it. In "On Sense and Reference" Frege had argued that the thought of a sentence could not be its reference because, if one substitutes a name in the sentence for another name, with the same reference but with another sense, the thought of the sentence is changed whereas its truth value remains the same. This argument, however, applies not only to the thought but to the complex ideas associated with the two sentences as well. Now, 26 years later Frege tries to close this gap by presenting an argument for the necessity of thoughts.

23. At first glance, it may seem strange that I roughly identify the realm of phenomena with Frege's sphere of sense. On second thought, however, it becomes clear that both are closely related insofar as the sense of a name is *"die Art des Gegebenseins"* [*des bezeichneten Gegenstandes*], that is, I guess, the way in which it objectively appears; an analogous proposal can be made for the expressions of concepts and relations: the sense of a concept-word is the property by which the truth-function is denoted.

24. Frege, "The Thought," *op. cit.*, pp. 23-24; the underlining is mine.

25. The relation between scientific thoughts, pictures, and phenomena can be expressed roughly in the following Tarski-like style: The thought of a correspondence between a picture and the phenomena is true iff the picture corresponds with the phenomena.

THE AESTHETICS OF THEORY TESTING: ECONOMY AND SIMPLICITY

Jane Duran

ECONOMY and simplicity are two aesthetic factors which play a recurring role in scientific explanation—or so we are told. At an early point in our acquaintance with philosophy of science we learn that, other things being equal, the simpler or more economical theory (in terms of hypotheses, auxiliary constructs, multiplication of entities, etc.) is to be preferred. We come to take such preference for granted and thus speak blithely and easily of the "simpler" theory. A short citation from some recent work on the notion of a theory may aid us here:

> [We can] compare how well theories are supported by evidence. [(1) one theory is to be preferred to a second if it does not contain hypotheses disconfirmed by a body of evidence, and the other does; (2) a theory containing fewer untested hypotheses is to be preferred to one containing more . . .] Such preferences are not founded, or rather need not be founded on the preference for better tested theories, and the various modes of comparison are only aspects of that preference . . . . I think there is rigor enough, however, to distinguish unambiguously among candidates that are offered . . . .[1]

Here we see that Clark Glymour seems to think that we can distinguish unambiguously between theories on a criterion of economy or simplicity, such that one theory containing fewer untested hypotheses is to be preferred over another.

One is tempted to think, at least initially, that Glymour is correct in this, and that what he has to say is more or less uncontroversial. If that were the case, a discussion of the notions of economy and simplicity might not prove very interesting. But I want to argue that what Glymour—or anyone else holding a similar thesis—has to say on this score is far from as intuitively clear as one might think. Economy and simplicity are difficult notions, I want to claim. It is by no means clear what counts as a more economical theory, and not simply because one cannot reach agreement on the constitutive elements of the theory (cannot reach agreement, say, on what constitutes a hypothesis within the theory). It is because these notions themselves are up for grabs, I shall argue, in a way parallel to the difficulties inherent in many constructs utilized in aesthetics *simpliciter*.

In order to make out my case I want to utilize some material from an older, somewhat disreputable portion of aesthetics: theory of beauty.

Theories of the beautiful had been out of fashion a long time when Guy Sircello published *A New Theory of Beauty* in the mid-'70's. Succinctly put, Sircello attempted to establish a set of necessary and sufficient conditions for the beautiful which, in terms of logical form, consisted of a biconditional with several conjuncts on the right-hand side. (In other words, an object is beautiful if and only if, etc.) The third conjunct reads as follows:

> (3) [A PQD of an "object" is beautiful if . . .] it is present in the "object" in a very high degree and any "object" that is not a PQD is beautiful only if it possesses, proximately or ultimately, at least one PQD present in the "object" to a very high degree.[2]

But it proves to be very difficult to ascertain what PQD (property of qualitative degree, as Sircello has it) is actually present in the object, if any, and to what degree the PQD is present. That this is so may not be surprising, but it is illustrative of the conceptual difficulty inherent in what at first promises to be a startlingly clear addition to the framework of the theories of beauty.

In order to make clear my parallel to the notions of economy and simplicity, I want to press the case with the difficulty of the straightforwardly aesthetic notion of PQD still further. Sircello makes much use in his examples of the property of vividness. In other words, according to the criterion presented above, vividness of a color (say, orange) is beautiful, according to the third conjunct on the right-hand side of this biconditional, if it is present to a very high degree. But Sircello's own examples show how difficult the construal of such a notion actually is. In order to provide an exemplar of vividness in color, Sircello alludes to sunsets:

> Most of us have no doubt had the experience of seeing a flaming (that is, extremely vivid) orange sunset, being struck by its beauty, and calling someone to share the sight only to have her appear after the color has waned to a ghost of its previous brilliance . . . . But the late arrival may not be disappointed at all, as she exclaims over the beautiful *bronze-gold* of the western sky. Having missed the vivid orange, she sees the very vivid bronze-gold, and hence a beauty of the sunset. And as the reddish tint gradually fades from the sky, a second latecomer may arrive and exclaim—and even get us and the first latecomer to exclaim—over the lovely *old gold* of the sky.[3]

This example is striking for its demarcation of the extent to which attribution of color, beyond the most general kind of categorizing, is dependent upon the percipient and also dependent upon fleeting qualities which are obviously time-bound. Sircello intends the example to be read as a case where the colors change, and the percipients recognize the various kinds of colors as they change, although it is clear that no one color lasts for a lengthy period of time. More to the point, however, the example can be read in another way (although Sircello does not do this): The first percipient's *orange* can be interpreted simultaneously by the second percipient as a *bronze-gold*, and so forth. Now all of this is intriguing because Sircello intends these examples to help clarify a PQD which one would think relatively obvious: vividness. Sircello's theory requires that this concept be made out, and yet it is clear that PQD's are too difficult to demarcate and not long-lived. Such recalcitrant cases would seem to constitute an obstacle to a well-argued theory of beauty; the colors change so quickly and are so far from resembling "standard" colors, both with respect to "vividness" and other properties such as saturation, hue, luminosity, and so on, that intersubjective agreement (which a theory of beauty would appear to require) seems to be at best a case of fortunate congruency.

In just the same way, I argue, it is difficult to make out the aesthetic notions of economy and simplicity of theories. Economy and simplicity are surely something more than *mere* ontological or hypothetical slenderness: In other words, when we speak of the "economy" or "simplicity" of a theory, we are probably referring to something more than a heavy-handed wielding of Occam's Razor. The situation is complicated by the fact that economy and simplicity are frequently mentioned in the same breath as elegance—although I will not attempt to address the status of that notion here, because surely it can (and should) be divorced from the other two conceptually.

The difficulty we encounter is, I believe, something like the following. Glymour is on the right track when he claims that we can distinguish unambiguously between theories on the number of untested hypotheses contained, and so on. But again, something *more* than such a distinction is required if we are to successfully make the claim that one theory is more economical than another, more appealing than another because of its simplicity, and so on. We are, also, I believe, implicitly speaking of a certain sort of formal structure. In other words, the more economical theory is not merely more slender (in a way analogous to my eight-year-old's having developed a better metaphysics because her ontology no longer contains such dubious entities as Santa Claus and the Easter Bunny), but ties its hypotheses together in a more appealing and more formally satisfying structure.

Theories may appeal to us in this way even when their ontologies (or sets of hypotheses) are bulkier. Our undergraduates, in moments of clear articulation, indicate that they prefer Plato's ontology to other slenderer ontologies because it is economical and simple in the sense that it—according to their worldview—ties everything together in a manner superior to some other, Occamish theories. Plato's metaphysics violates the Razor, as Aristotle was quick to see, because it introduces unneeded entities which have, on an empirical view, little or no explanatory power. And yet Plato's metaphysics appeals to most of us on a level which, I claim, Aristotle's does not. Which theory is more "simple"? Which theory is more "economical"? There are many respects in which one is tempted to cite Plato's view.

Work in twentieth century physics seems to provide us with yet another example of the difficulty in attribution of aesthetic constructs to scientific theories. One does not need a sophisticated grasp of mathematics or physics to achieve a bare-bones understanding of the move from a Newtonian worldview to, say, the worldview of quantum mechanics. In undergraduate courses for those in the liberal arts, we are told that the Newtonian worldview contained notions of absolute space and time, of the inherent assignment of an entity to a certain designated point in space or time, and so forth. In other words, the physics of the seventeenth century contained a certain sort of simplicity and elegance, and even on crude Occamish grounds, arising as it did from the work of Copernicus and Galileo, was more ontologically thin than many of the preceding views. But then, we are told, quantum mechanics and relativistic spacetime appeared, and the simplicity of a notion such as Absolute Space was destroyed, along with certain principles of inherent measurability, at least on the level of particles. Again, one ponders the rhetorical question implicit here: it is by no means an easy task to decide which view is more economical. Quantum mechanics ridded us of notions which were both unnecessary and lacking in the requisite explanatory power—in its place we are given a certain implicit undecidability, which is perhaps not as appealing aesthetically.

The difficulties with economy and simplicity, then, mimic or remind us of difficulties we have with other concepts useful in aesthetics or found in aesthetic contexts. As we have seen, an analogy can be made between an alleged property of qualitative degree, such as vividness (which according to a theory of beauty inheres in an object), and the notion of economy. Interestingly enough, the tables can be turned the other way, and an analogy can also be constructed between an aesthetic notion such as expression—which is an element of intentionality, and does not reside in the object under consideration—and a notion such as economy or simplicity.

In a recent paper I argued that Collingwood's expressionism was noteworthy not only for its reliance on intentionality, but for its failure to distinguish between levels of intentionality.[4] Thus I argued that part of the solution to the problems posed by Collingwood's art/magic distinction lies in some of what the commentators, notably Armstrong, have had to say about the art/craft distinction: ". . . [while] art does differ from craft, it differs not by being wholly other than craft but by being more than simple craft."[5] Analogous remarks might be made about the use of the term "economical" in theoretical contexts, although admittedly the analogy is more difficult to make out.

To call a theory more "economical" than its competitors is not, in general, *merely* to say, for example, that the theory has fewer untested hypotheses. (Here the analogy breaks down a bit, because we have no parallel term for a theory that does merely contain fewer untested hypotheses—perhaps we should want to call such a theory "skinnier," or "slenderer.") A theory is more economical than its competitors when it is slenderer, and, typically, when it is more satisfying in some other sort of way, which may, in fact, be difficult to articulate. Aristotle's ontology may contain fewer entities, but we may well want to think of his ontology as merely "thinner," not more "economical" or "simpler" than Plato's. In fact, so far as the dubious term "simpler" is concerned, it may very well be Plato's ontology which we want to label as such.

The element of intentionality here is as follows: the terms "economy" and "simplicity" have difficulties on an attributive level similar to those of, say, vividness. When we apply such a term, there is a large intentional component of unarticulated beliefs and clusters of semantic intension which yields the net result that the application of the term is much more a construct of our intentions than is, for example, the application of the more straightforward "horse," "dog," or "refrigerator." The intentionality is not itself, of course, completely or plainly expressionistic, as is the intentionality which Collingwood thinks is the hallmark of a work of art. But it is tinged with such an element, and in addition to that particular difficulty, it is clear that insofar as there is a notion of *economy* which has Fregean sense and connotation, it does not have such connotation in any straightforward kind of way. The application of these notions—economy and simplicity—is then a difficult and recondite matter, one itself worthy of serious study in aesthetics.

To conclude, I have argued that a notion such as economy, when appended to scientific theories and/or constructs, initially strikes us as being trivial and uninteresting. A first gloss is that a theory is more economical than another when it contains fewer untested hypotheses or when it posits fewer entities than another. (Glymour was cited in this regard.) But a second look forces us to the conclusion that it is by no means clear what is generally intended by the application of these terms, or when they are applied. Sircello's *A New Theory of Beauty* was cited, and an analogy constructed between the recognition of a property of qualitative degree (such as vividness of a certain color) and recognition of the economy or simplicity of a theory. It was then noted that thinness of ontology is apparently not, in many cases, what distinguishes a more "economical theory" (Plato vs. Aristotle), and the move from classical mechanics to quantum mechanics was alluded to. Finally it was noted that there is a strong intentional component to the use of such terms which, I contended, is weakly analogous to some of the conceptual difficulties I have recently cited in Collingwood.

Philosophy of science, then, has an intersection with aesthetics. The fact that this intersection is largely unremarked upon and indeed, even unnoticed, does not mean that it does not contain fertile ground for conceptual analysis.

University of California at Santa Barbara

NOTES

1. Glymour, Clark, "The Epistemology of Geometry," *Nous*, vol. 11 (1977), pp. 234-35 and *passim*.

2. Sircello, Guy, *A New Theory of Beauty* (Princeton: Princeton University Press, 1975), p. 43.

3. *Ibid.*, p. 26.

4. See my "Collingwood and Intentionality," in *British Journal of Aesthetics*, vol. 27 (1987), pp. 32-38.

5. Armstrong, A. MacC., "The Primrose Path to Philistinism," *British Journal of Aesthetics*, vol. 23 (1983), p. 146.

SIMPLICITY IN EVOLUTIONARY EXPLANATIONS

David B. Resnik

EVOLUTIONARY biologists appeal to a variety of epistemic values in explanations of organic evolution. Some of these frequently used "aesthetic" parameters include simplicity, parsimony, and explanatory power. In this paper I will examine simplicity's role as a parameter in evolutionary explanations. I will argue that at this point in time at least, biologists should not use simplicity as a reason for favoring one evolutionary hypothesis over another, other things being equal.[1] My reasons for attacking the use of simplicity in evolutionary explanations are twofold. First, evolutionary biologists do not now have a method for determining what makes one evolutionary hypothesis simpler than another, and no such method may be forthcoming. Second, even if evolutionary biologists could develop a way to rank hypotheses according to degrees of simplicity, they would still not have good reasons for believing that the simplest explanation of an evolutionary process is most likely to be true (or most acceptable), other things being equal.[2] I shall defend my two reasons for attacking the use of simplicity as an "aesthetic" parameter in evolutionary explanations, and I will suggest simplicity still has an important use in evolutionary biology as a heuristic principle.

I. Simplicity: How Simple?

Scientists and philosophers use the term "simplicity" without a clear idea of what it means. We make claims that one hypothesis is simpler than another, and we often readily assent to such claims. Yet when asked why one hypothesis is simpler than another, we are often at a loss for words. Our ordinary and even sophisticated use of simplicity is often vague and ill-formed. Evolutionary biologists who appeal to a principle of simplicity in confirming hypotheses must go beyond our ordinary, vague use of simplicity: they must be able to show why a given hypothesis is simpler than another hypothesis. Furthermore, if they wish to employ simplicity as an epistemic value, they must be able to give good reasons for thinking that simpler hypotheses are better than their more complicated competitors.

What criteria might an evolutionary biologist use to determine the simplest hypothesis? One possible approach would be to analyze simplicity in ontological terms: the simplest hypothesis is the hypothesis that commits one to the existence of the fewest entities. This approach, though quite useful in other sciences, is not very useful in evolutionary biology. Any explanation of organic evolution requires a whole zoo of biological entities including organisms, populations of organisms, phenotypic traits, genomes, organ systems, and organelles. In addition, in order to explain the operation of biological mechanisms and processes, an evolutionary biologist must also assume the existence of a myriad of nonbiological entities including ecosystems, rivers, oceans, clouds, the sun, the earth, molecules, atoms, and untold subatomic particles. Biologists might object that we could formulate a principle of simplicity in terms of biological entities, but this move would still not simplify the problem. It will still be difficult, if not impossible, to enumerate and distinguish between all the different entities using any ontological notion of simplicity. Thus the ontological approach does not give evolutionary biologists an effective or useful notion of simplicity.

Instead of analyzing simplicity in terms of ontological commitments, perhaps biologists might find it more useful to adopt a principle of simplicity which gives preference to simple processes or mechanisms. It would seem that anyone can tell the difference between a simple and a complex process or mechanism: a lever is a simple mechanism; a computer is a complicated mechanism. Likewise, combustion is a simple process, but learning is a complicated process. Although we seem to be able to distinguish between simple and complex processes or mechanisms in these everyday cases, our intuitions may fail us when we try to make the same distinctions in evolutionary biology. For example, consider the following set of possible explanations for the evolution of altruism. Altruism could be due to group selection, kin selection, individual selection, or random drift (no selection at all). Group selection for altruism requires that a given population of organisms has a number of groups which have differential reproduction between groups, and there must be some restrictions on gene flow between the groups. The kin selection hypothesis requires that altruists help mostly their close kin and that altruistic individuals enhance their inclusive fitness. The individual selection hypothesis requires that altruists derive some direct benefits from their altruistic acts, such as protection from predators or experience in rearing young. The random drift hypothesis would seem to be false, since altruism usually decreases individual fitness and should be selected against. At the very least, the hypothesis that random drift is responsible for the evolution of altruism would require a tremendous number of coincidences. All of these possible explanations of the evolution of altruism postulate elaborate mechanisms or processes. In addition to evolutionary processes, these explanations also postulate a number of phylogenetic, ecological, behavioral, and molecular processes. It is not at all obvious which mechanism or process is the simplest one. In cases like these it will be very difficult to decide which hypothesis is the simplest hypothesis, if we use a notion of simplicity which is based on an analysis of processes or mechanisms. Perhaps as our knowledge of the evolution of life on this planet grows, we will develop a clearer understanding of what makes one biological process simpler than another one, but that time has not yet arrived. I suggest that evolutionary biologists do not at this time have a clear notion of what constitutes the simplest evolutionary process or mechanism.

If one approach fails, try another. Perhaps a theoretical approach might prove the most useful. An evolutionary biologist could maintain that the simplest hypothesis is the one that uses the simplest theory. In order to avoid circularity, one could maintain that the simplest theory is the one that uses the fewest axioms, postulates, and laws. But again, I think that this approach will not prove useful. There are often many different ways to formulate any given theory; one version of a theory may have many axioms; another version may have but a few. For example, philosophers and biologists have formulated the theory of natural selection in a number of different ways. According to Mary Williams' version of the theory it has five different axioms; according to John Beatty's version it has only two.[3] A theory which seems simple under one formulation may not appear to be so simple on another formulation. As long as there are many different and equally useful formulations of a given theory, biologists will have to choose between these different formulations. Supposing that evolutionary biologists find a simplest formulation or a simplest theory, it is not at all clear that the simplest formulation of a theory or the simplest theory will be better than its competitors, other things being equal. Biologists need a reason for preferring the simpler theories. A principle of simplicity, by tself, is not sufficient justification for accepting a theory or an hypothesis.

There are other kinds of simplicity which evolutionary biologists might use. For example, they might try to analyze simplicity in mathematical, logical, or even strictly practical terms. I do not pretend to have exhausted the list of possible ways to analyze simplicity.[4] Regardless of how biologists might analyze this notion, it is clear that they do not now have a commonly accepted definition of the term or how it ought to be applied in evolutionary biology. Perhaps they may one day develop an effective notion of simplicity in evolutionary explanations, but I suggest that that day will not arrive soon, and perhaps will never arrive.

II. Is Simpler Better?

Let's suppose that evolutionary biologists have managed to agree on an analysis of simplicity. It would still be an open question whether the simpler hypotheses or theories are better (or more likely to be true) than their competitors, other things being equal. I would like to suggest that in evolutionary biology, at least, the simpler hypotheses or theories will not be better or more likely to be true. Organic evolution is often a very complicated process involving many different selection pressures, chance events in the environment, interactions between organisms and their environment, and interactions within the genome. Almost any phenotypic trait is the product of many different mechanisms and processes. Indeed, given the complexity of the evolutionary process, we have good reasons for believing that the "simplest" explanation of evolution is likely to be false.

Consider, for example, the well-documented and well-known case of industrial melanism in moths.[5] During the nineteenth century in Liverpool, England, air pollution killed most of the lichens that grew on tree trunks. The peppered moth (*Biston betularia*) had used these lichens for camouflage to escape predation by birds. As the trees lost their light-colored covering, a darker form of the moth appeared and became the most common form of the moth. The best explanation of the evolution of these moths is that these moths evolved according to natural selection: the dark form of the moth survived because it was better adapted to the changed environment. The dark moths were better camouflaged, and thus better able to survive and reproduce than the light colored moths in the changed environment. The dark moth's color was an adaptation to dark, lichenless trees. Industrial melanism is a classic example of a good explanation of organic evolution, but is the industrial melanism hypothesis simpler than other possible hypotheses? It might seem like a straightforward, simple hypothesis, but on closer examination we can see that it is not. It commits one to the existence of all sorts of biological and nonbiological entities, such as moths, lichens, trees, moth wings, moth wing pigments, birds, pollution, polluters, and so on. It postulates the existence of a number of elaborate mechanisms, such as a mechanism for producing wing pigments, and mechanisms for passing on hereditary traits. It also uses a complicated theory, the theory of natural selection. So according to all the notions of simplicity examined in this paper, the industrial melanism hypothesis is not a simple hypothesis, and it might not be the simplest explanation of industrial melanism. The hypothesis that industrial melanism is caused by random drift might commit biologists to the existence of fewer entities, invoke "simpler" processes, and use a "simpler" theory. But it would not be the best explanation of industrial melanism. The best (commonly accepted) explanation of industrial melanism is not simple nor is it simpler than other explanations.

At this point someone might object that my discussion of this example presupposes a clear notion of simplicity. In claiming that the industrial melanism hypothesis is more complicated than other possible competing hypotheses, I have assumed that we can tell what makes one hypothesis simpler than another. My discussion is unfounded and pointless without a working notion of simplicity, according to this objection. But I do not need to have a working notion of simplicity or a definition of the term in order to make my point. I am merely appealing to our intuitions about simplicity. While they may be still vague and ill-formed, I maintain that the industrial melanism hypothesis should be considered a complicated hypothesis no matter how we analyze simplicity. Again, someone might argue that once we develop a good theory of simplicity, we might see that the industrial melanism hypothesis is in fact simpler than other hypotheses. So the dispute boils down to this: what comes first, intuitions or theories? Should we develop a notion of simplicity and then apply it to examples or should we form a notion of simplicity based on some clear-cut examples? I don't have much to add to this kind of dispute, and I certainly won't say much more about it in this paper.

Someone else might object that I have presented a biased sample. The explanation of industrial melanism in moths might not be the simplest explanation, most good explanations of evolution are the simplest explanations. While I agree that sometimes the simplest explanation will be the best explanation, I do not think it is the case that the simplest explanations will always or even often be the best explanations of evolution. Indeed, I chose the industrial melanism example because it seems like a *prima facie* case of a simple evolutionary explanation. However, I think on closer examination it turns out to be not so simple, and certainly not the simplest possible explanation. I maintain that most of the apparently "simple" explanations may turn out in fact to be quite complicated. We could of course discuss examples until we have surveyed all the explanations in evolutionary biology. But I will not undertake that project here.

III. What's Good About Simplicity?

If I have argued correctly up to this point, then it would appear that evolutionary biologists do not have much use for a principle of simplicity in evaluating evolutionary explanations. They do not have an effective notion of simplicity and none may be forthcoming. Even if they did have a clear understanding of simplicity, simpler hypotheses are not necessarily better or more likely to be true. So what role, if any, ought simplicity play in evolutionary theory? I think simplicity can still plan an important role in evolutionary theory as a heuristic principle. One could state the principle as follows: "First formulate and test the simpler hypotheses, and if these fail, test more complicated ones." Evolutionary biologists can use a principle of simplicity to suggest possible hypotheses and guide research. Simplicity may not lead evolutionary biologists to the truth or to better explanations, but at least it can yield manageable hypotheses which serve as a starting point for research and suggest further hypotheses. The principle may not play a crucial role in confirmation or explanation, but it can still play an important role in discovery. Moreover, biologists do not need to have a clear or rigorous analysis of simplicity in evolutionary explanations in order to use simplicity as a heuristic device. An informal notion of simplicity is sufficient to guide evolutionary research. For instance, one might try to characterize simplicity in terms of economy of cognitive means.[6]

Biologists could spurn simplicity as a heuristic and adopt an opposing principle instead, but I think such a principle would not be a useful heuristic. Suppose they adopt a principle of complexity instead of a principle of simplicity. In following this principle, they would formulate and test the most complicated hypotheses first. This principle would not prove fruitful because it would not yield manageable hypotheses nor would it allow biologists to easily pursue alternative hypotheses. A principle of complexity might occasionally lead biologists to true or acceptable explanations, but it would also make evolutionary research more difficult. Given the choice between simplicity and its opposing principle as heuristics, simplicity is the better heuristic.

Perhaps simplicity will one day play an important role in evolutionary explanations; it remains to be seen whether biologists do develop an effective understanding of simplicity and how it applies to evolutionary explanations. For now at least, simplicity should be regarded as an "aesthetic" parameter in explanation and nothing more. It may help us create explanations that are more beautiful, manageable or pleasing, but it does not necessarily produce explanations that are likely to be acceptable or true.[7]

University of North Carolina at Chapel Hill

NOTES

1. The use of the phrase "other things being equal" is a bit misleading, since we almost always have some reasons (empirical or methodological) for preferring one hypothesis over its competitors. When I use this phrase I am supposing that the competing hypotheses are equal except with regard to their alleged degree of simplicity.

2. I use the phrase "true or most acceptable" here in order to remain neutral on questions of realism in science. I think my views apply to both realist and non-realist interpretations of science.

3. See M. B. Williams, "Deducing the Consequences of Evolution: A Mathematical Model," *Journal of Theoretical Biology*, vol. 29 (1970), pp. 343-85, for her version of the theory and John Beatty, "What's Wrong with the Received View of Evolutionary Theory?," *Philosophy of Science Association*, vol. 2 (1981), pp. 397-426, for his version of the theory.

4. For more on the analysis of simplicity see Elliott Sober's *Simplicity* (Oxford: Clarendon Press, 1975).

5. H. B. D. Kettlewell, "Selection Experiments on Industrial Melanism in Lepidoptera," *Heredity*, vol. 9 (1955), pp. 323-42.

6. Nicholas Rescher argues for a cost-effective or economic approach to simplicity. See his paper in this volume.

7. I would like to thank Tom Adajian, Robert Brandon, William Lycan, Michael Resnik, Jay Rosenberg, Brad Wilson, and Paul Ziff for their helpful comments and criticism.

THE ROLE OF SIMPLICITY IN GEOLOGY

William R. Brice

IDEAS and concepts created to explain natural phenomena will frequently appear to be unrelated and isolated when first developed. As these ideas undergo further scrutiny, many must be abandoned or modified when they fail to explain the observable facts. Eventually, when the real forward steps are made, larger elegant, simple, all encompassing ideas either replace or incorporate what has gone on before. But in all of this, the best working hypotheses generally have been the ones that were the simplest. Such changes have occurred many times in geology, and two of the best illustrations came almost 200 years apart. The first is the "Neptunists" concept of Werner in the eighteenth century and the second is the current revolution of Plate Tectonics.

I. Introduction

The modern science of geology seemed to suddenly burst forth in the intellectual world during the latter part of the 1700s and early 1800s. But geology, as with all sciences, was built on a body of previous knowledge and the advances that occurred when researchers were able to break with the past and forge new directions for the old facts. This "late blooming" of geology must not be interpreted as indicating that geology is an inferior science, but rather reflects how dependent geology is on the other sciences. The understanding of the Earth could not really move out of the realm of myth and fantasy until the basic understanding of other sciences such as chemistry, physics, and biology was better developed. Because of this dependence, concepts in geology can never be developed to the point where they would become independent of the other sciences. Porter (1977) and Green (1982) provide marvelous overviews and analyses of this early development of geology.

The idea of simplicity in science is generally described as the application of Occam's razor or the principle of parsimony. William of Occam, who is known also as "Doctor Invincibilis" and "Venerabilis Inceptor," was a Franciscan monk who lived in the early fourteenth century. The original wording used by Occam is somewhat irrelevant (see Laird 1919), but the principle has become widely used. One of my favorite interpretations is given by Russell (1929), in which he said that entities should not be multiplied beyond necessity. Over the years there has been a great debate as to whether simplicity is an artifact of the human endeavor to explain nature or whether simplicity is the result of our assumption that it is inherent in nature (see Kemeny 1953). Feuer (1957), following in the direction set by Mill almost 100 years before, felt that simplicity when applied in scientific terms is not related to any assumption about nature, but rather it deals with methodology of science. In Feuer's view, verifiability becomes the dominant aspect, and the unverifiable items are disregarded as unnecessary. This interpretation may well be a way of "playing it safe" and only working with ideas that match the greatest number of observed facts.

An alternative was suggested by Thomas Chamberlin (1834-1928), former President of the University of Wisconsin and Professor at the University of Chicago, in his well known paper entitled "The Method of Multiple Working Hypotheses" (Chamberlin 1890). Chamberlin argues that it is better to keep several competing hypotheses in mind while investigating a set of facts or doing field work, rather than to begin with a preconceived notion of what one hopes to find:

> The moment one has offered an original explanation for a phenomenon which seems satisfactory, [at] that moment affection for his intellectual child springs into existence; . . . There is an unconscious selection and magnifying of the phenomena that fall into harmony with the theory and support it, and an unconscious neglect of those that fail of coincidence. (Chamberlin 1890.)

Although Chamberlin does admit that such an approach carried too far can lead to vacillation and the eventual failure to come to any decision, he concludes that with certain limitations, such an approach is preferable. Thus, even within Chamberlin's multiple hypotheses, the rule of simplicity and parsimony seems to work best. However, when applied without sound physical evidence in support, simplicity can be both a vice as well as a virtue.

II. Werner

Historians seem to agree that geology, as it is understood today, had its origins in the years between about 1750 and 1850, and although there were many who contributed to this initial development, Abraham Gottlob Werner (1749-1817) is preeminent.

Werner, born in Prussian Silesia on September 25, 1749, was raised around minerals and geology, for his father was a manager of mine works. He enrolled at the Bergakademie at Freiberg in 1769. He quickly exhausted the course of study there and went on to Leipzig in 1771 where he stayed until 1774. At Leipzig Werner began studying law, but changed to languages and mineralogy, and while still a student at Leipzig he published his first book in 1774, *On The External Characteristics of Fossils* [*Von den Aeuszerlichen Kennzeichen der Foszilien*], using the original definition of "fossil," namely anything that was dug from the Earth. The article was really about what today we would call minerals. This work was well received by the scholars and greatly enhanced his reputation. Although he was offered positions in eight different countries, including France, Poland, England, and Austria, he chose to stay in his homeland. By the age of 25 he was a mine inspector and teacher of mining and mineralogy at the Freiberg Academy, where he practiced the art of observation. It was Werner's contention that students must know how rocks and minerals look in place down in the mines, not just in the collection drawers. He stayed at the academy in Freiberg until his death. For 42 years, despite numerous offers to go elsewhere, he stayed. His salary rose to over 3000 thalers at a time when skilled workers in the mines were earning about 60 thalers.

He did very little publishing, and even his lecture notes were seldom more than a few words scribbled on a slip of paper. What we know of his ideas comes more from the writings of his students than from the master himself, for Werner became one of the great teachers of his time, whose fame was such that people would master the German language in order to attend his lectures. He was extremely methodical, and it was this capacity for orderliness that helped him create order out of the geological chaos he saw around him, an order that few before him had been able to see. [For more details on Werner's life see Cuvier's (1819) near-contemporary account and other shorter descriptions in Fenton and Fenton (1945), Geikie (1905), Zittel (1901), and Ospovat (1971).]

Many of the theories of the Earth and its geology existing at the time of Werner were vague and grandiose, having religious overtones, and practically all of them were constrained by a time frame put forth in 1650 by Bishop James Ussher. He proposed that creation occurred the evening BEFORE Sunday, October 23, 4004 B.C., a date that most people misquote (Brice 1982). Thus, any idea, if it was to be accepted, had to fit within Ussher's date of creation and not contradict the Bible, and Werner's work reflected this bias.

III. The Simple Idea of Werner

Although Werner published very little, we do have a record of his major contribution, known in English as his *Short Classification and Description of the Different Rocks* [*Kurze Klassifikation und Beschreibung der verschiedenen Gebirgsarten*], which was published as an article in 1786 and then in pamphlet editions in 1787 (Ospovat 1971). This 28 page pamphlet outlines Werner's concept for the whole Earth. Even though, as I shall demonstrate, he tended to be too simplistic and dogmatic for reality, he did provide a unifying paradigm, one of the first to be seen in geology. Also, for the first time the classification of rocks was separated from mineral classification.

Werner's classification is the very model of what is now called a stratigraphic sequence that starts with the oldest rocks and moves to the youngest. Werner was not the first to attempt a subdivision of the rock strata, and no doubt he was influenced by the work of such people as Giovanni Arduino (1713-1795), a Tuscan mining engineer, and John Strachey, who between 1719 and 1725 contributed papers in which he enumerated, in proper sequence, the geological formations in southwest England. In Germany Johann Gottlob Lehmann (d. 1767) and George Christian Fuchsel (1722-1773) both published material with which Werner would have been familiar (Geikie 1905). According to Green (1982), Werner appears to have been strongly influenced by "A Physical Description of the Globe," published in 1769 by a Swedish writer, Tobern Bergman (1735-1784). But Werner's was the first classification that attempted to present a complete system for the Earth, including the origin of the rocks.

Steno, working in northern Italy about 100 years earlier, had already invented the idea of "Superposition," that is, that the oldest rock is on bottom and the youngest is on top, but no one had attempted to apply it in such sweeping terms. Werner based his classification upon the very popular idea known as the "Neptunist" theory; a primitive ocean [the Great Flood?] surrounded the globe and eventually precipitated the "primitive" rocks covering the uneven core forming the original crust. [In today's terms, Werner's primitive rocks are the granites and crystalline gneisses seen in many mountain ranges.] As a result of this precipitation, the level of the ocean dropped enough to allow some high mountain tops to be above water. This explains why crystalline rocks are found on some mountain summits and are not covered by the succeeding rocks. Later Werner added an intermediate stage after the primitive epoch, which he called the "transitional" rocks. These were partly precipitated and partly formed by mechanical action, and they provided barriers that subdivided the surface into "land" and "ocean" areas and marked the end of the universality of this original primitive ocean. [Many schists and similar metamorphic rocks fall into Werner's transition classification.] Violent storms and turbulence then broke up and eroded the earlier material and formed the totally "mechanical" rocks, or Floetz rocks, as Werner named them. [These would be what we know today as the sedimentary rocks.] Even though the major ocean was gone by this time, there were, in Werner's concept, still some local flooding events [not much different from the modern theories of marine transgression and regression]. Werner felt that the alluvial and volcanic materials were all due to local events. Also, Werner was convinced that the universal ocean would not have been uniform in its composition either in location or time; thus he could explain the variation seen in rock strata of the same position and similar rocks found in different times. Using the results of a very timely discovery that water is made of two gases, hydrogen and oxygen, Werner overcame the problem of what happened to the water by having it simply decompose into its gases and be used to make the atmosphere. Although he never did fully account for the source of the water in the first place or the dissolved materials which formed the rocks, he felt that the inability to explain all phenomena should not preclude recognition of their existence and the conclusions drawn from them. Werner said, "In all researches into effects and their proximate and remote causes, we arrive at least at the investigation of ultimate causes, beyond which we cannot proceed" (Ospovat 1971). But ultimate causes or not, with this single concept and classification, Werner could account for most of the geologic formations observable in his day, except for certain volcanic rocks which finally proved fatal to his ideas and helped bring about the Wernerian downfall. But until that happened, it was truly a universal concept.

In time, Werner's concept failed for several reasons, not the least of which was that he simply did not have a large enough data base from which to draw his conclusions. Even though he had travelled to Paris and the Alps, his body of direct field knowledge was confined to the region of central Germany where he was born and lived his entire life, and he can be criticized for over-generalizing from a limited field of knowledge. However, the stratigraphic relationships which he determined by the application of Steno's idea of Superposition are, for the most part, quite correct. Modern analyses of these strata to some extent tend to be only a refinement of Werner's original work (Ospovat 1969). Indirectly through his reading, Werner was acquainted with the work of his predecessors and some of his contemporaries, and he made use of their data as well. However, the problem seems to be that he tended to select and believe the data that supported his idea and not give much thought to the parts that did not fit, such as the true place of volcanic rocks, especially basalt. Werner appears to illustrate the situation mentioned by Chamberlin where he was too enamored with his own idea to see the importance of data that ran counter to it.

Werner's concept became the starting place for much of the subsequent development of geology. People were drawn to it because of its simplicity and because in general applications his ideas did really seem to work. He had a mechanism to make rocks by using both precipitation from the universal ocean and mechanical processes; there was an order to the precipitation and subsequent reworking which explained the general order of strata seen in the field. By using both precipitation and mechanical processes, Werner explained why there are different kinds of rocks, and this was very attractive. Even the early work by William MacClure in the United States used Werner's system of rock classification. Many people were attracted to Werner's idea because of its simplicity, its apparent universality, and because it fit within the Biblical framework, that is, the Great Flood. However, as flaws in the system began to turn up people began to move away from the Wernerian ideas. Adam Sedgwick, the giant of British stratigraphy in the early 1800's is a good example of one who started with Wernerian beliefs and then changed. He is reported to have said, ". . . for a long while I was troubled with water on the brain, but light and heat have completely dissipated it," and he eventually referred to, ". . . the Wernerian nonsense I learnt in my youth" (Clarke and Hughes 1890). Eventually the counter evidence overwhelmed the Wernerians, and Werner's concepts were almost totally replaced, at least the theoretical portions. As mentioned above, his stratigraphic relationships were quite good. The concept of using heat and rock melting, which replaced that of ocean origin for the crystalline rocks, came from another eighteenth century worker, James Hutton, whose ideas seem to have been closer to the mark than Werner's were. As a result, Hutton rather than Werner is known as the "father of modern geology" (see Bailey 1967 and Hutton 1788). However, Werner holds a prominent place because he started geology in the right direction by producing an idea which provided a starting point and inspiration for others.

IV. Twentieth Century Simplicity

The development of the current revolution in geology is much more difficult to trace back to a single person than was the Wernerian revolution of the eighteenth century. This is partly because the body of knowledge on which it is based is so much larger than the one Werner used, and partly because more people were involved in basic research. The development of the current concept started with the ideas of continental drift, an "outrageous" hypothesis that began surfacing almost at the same time the followers of Hutton and Werner were arguing the relative merits of water vs. heat. Even Francis Bacon as early as 1620 noticed the similarities of the shape of the coastlines of western Africa and eastern South America, but little else was done with the idea for over 200 years.

Antonio Snider (1858) appears to be the first to actually suggest that the continents actually moved, and in keeping with the religious bias of that time, he suggested the cause for the movement was the Biblical flood. Most people considered the idea of moving continents as something from the lunatic fringe. Even if the idea was acceptable, they disagreed with this catastrophic approach for moving the land masses. A few, like Pickering (1924), preferred to relate the movement to the formation of the moon out of what is now the Pacific Ocean basin. However, the early approaches had little factual support other than the similar shapes along the South Atlantic coasts. During the latter part of the nineteenth century as geology grew as a science, field workers began to notice similarities in rock type, rock sequence, and fossils, especially in the continents of South America, Australia, Africa, and India, known as the "Gondwana Land" continents. In particular, the fossils presented an almost insurmountable problem, for many of the plants and animals could not have survived in salt water and were too heavy to have been blown by the winds from one continent to another.

Continental drift eventually became the only plausible answer, because it was the simplest idea which fitted the facts at hand, outrageous as it seemed. The alternatives of interconnecting continents or land bridges between the existing land masses were fully explored, but these ideas could not overcome the great differences that exist between the rocks of the ocean floor and the continents. The idea of land bridges was the last alternative to really be put to rest, but not before the large bridges had been reduced to narrow isthmuses and eventually to a series of islands, with rafting of organic material from island to island. Even as late as 1952, a few people were still advocating some kind of land bridge (Mayr 1952). Eventually ancient climate patterns and extensive evidence of almost simultaneous glaciation in the Gondwana continents during the Carboniferious Period (approximately 360-300 million years ago), left continental drift as the only solution. Probably the best known proponents of continental drift of the early twentieth century are Alfred Wegener of Germany and Frank B. Taylor of the United States (Wegener 1912 and 1924, and Taylor 1910). To say that Wegener's idea was greeted with scorn would be putting it too mildly. The established geologists, especially in North America, would barely give him a hearing, and Taylor's paper fared little better after its presentation at a meeting of the Geological Society of America in 1909. Critics exploited Wegener's lack of a mechanism to cause the continents to move and the physical difficulties of having these great continental masses plow their way across the ocean floor. However, in some cases, as geologists began to attempt to unravel such complicated areas as the Alps, people such as Collet (1926) were left with the inescapable conclusion that Wegener's idea of moving the continents was the only one that fitted the facts seen in the field. After describing the geology and even recognizing that part of Africa was actually resting on Europe, Collet could account for what he saw ONLY by having Africa collide with Europe. He said:

> I must state explicitly that all these results [the field geology and mapping] have been obtained independently of Wegener's hypothesis. That is why I think that they are a great support to Wegener's theory . . . . it will be soon impossible to resist Wegener's attractive idea. (Collet 1926.)

Regardless of the beauty and simplicity of continental drift and the limited factual evidence put forth, most people felt that it still belonged to the lunatic fringe. The only place the idea received much support was in the southern hemisphere, which is where most of the evidence was found prior to World War II (Du Toit 1937).

During the period between the two great wars and immediately afterward, more and more unexplainable phenomena were discovered, for example, the location of deep-focus earthquakes (Wadati 1935), and the structure of the ocean floor (Raitt 1949, and Tolstoy and Ewing 1949), that, unlike Collet's description of the Alps, did not fit into the elegant idea of moving the continents across the ocean floor. Like Werner's concept before, Wegener's idea was running into renewed difficulty as new and more detailed knowledge about the Earth came to light. A new synthesis was needed.

The major building blocks for this new synthesis came primarily from two areas of development, paleomagnetism and seismology, the study of earthquakes. Paleomagnetism, the study of ancient magnetic fields preserved in crystalline igneous rocks, indicated that the magnetic polarity had reversed itself many times in the past (Cox *et al.* 1963) and that the magnetic poles for Europe and North America had not always been stable, apparent polar wandering it is called ("apparent" because there is no evidence to suggest that the magnetic poles and the geographic poles have not always been in close proximity to each other and in the same location relative to the Earth as a whole). Furthermore, each continent produced its own polar wandering curve with the land masses in their present position. Only in the last few million years did the two curves coincide (Runcorn 1956). Two alternatives were possible here: either in the past the Earth had two distinct magnetic field systems, one for North America and one for Europe, or the continents had not always been in their present position, that is, they have drifted. Given the physical implausibility of two different magnetic field systems, the less outrageous hypothesis of continental drift seemed more attractive, and Runcorn postulated a drift of North America of about 24 degrees westward relative to Europe in the last 200 million years, just about what Wegener had predicted. But there was still the problem of having no known driving mechanism for the drift and no answer as to how the continents slid over the ocean floor without causing vast deformation.

Another important contribution concerning paleomagnetism came from the work of Vine and Matthews (1963) and their descriptions of magnetic anomalies found at the mid-ocean ridges. They found similar patterns of positive and negative anomalies on either side of the ocean ridges, and they created a model which would explain these observed patterns. However, this new model did not fit Wegener's concept, for Vine and Matthews had to assume an impermanence for the ocean floor, as suggested by Hess (1962) and Dietz (1961), which Wegener did not do. For Wegener, the ocean floor was permanent and did not move.

Seismology was to produce the other key ingredient for the new synthesis. Several people, but especially the group at Lamont-Doherty Geophysical Observatory, were studying the locations, depths, and first-motions of earthquakes. Out of this work came the idea of the crust being subdivided into several large plates of solid rock material, called lithosphere, made primarily of ocean crust, with some plates having masses of continental material sitting on them. These plates, in turn, are "floating" on a partially melted zone of the upper mantle. The plate idea, coupled with sea-floor spreading and the location of deep-focus earthquakes, eventually led to the concept of so-called "plate tectonics" (see Isacks *et al.* 1968 for a good summary).

Finally, in the late 1960's, geology, once again, had a paradigm that had world-wide application and could explain many seemingly unrelated phenomena. Here was a mechanism for moving the continents; they just ride on the large plates like leaves on a stream, without any control over their own destiny. The continents move as the plates move, and when continents collide, the lighter continental material will not go down into the mantle, but rather it tends to crumple and fold, producing large mountain ranges, such as the Himalayas and the Appalachians. This concept is applicable on a scale only dreamed of by Werner, and certainly goes far beyond the limits of Wegener's continental drift.

However, while this concept is beautiful in its simplicity and elegance, it cannot explain everything. Although it comes close to being the "Unified Field Theory" of geology, there are still problems to be solved. The concept of plate tectonics, too, suffers from the same malady that confronted Wegener, namely a mechanism to cause the plates to move. The moving plates can cause the continents to drift, but that only begs the question, and does not tell us how to move the plates. Perhaps it is density differences between the plate and the underlying asthenosphere, as Press (1969) suggests; perhaps it is heat flow from the interior of the Earth (Morgan 1971), or perhaps it is something we have yet to discover. As Werner's concept left unanswered questions, so does the concept of plate tectonics.

In the initial stages of applying this concept, it seemed to offer the answer to an age-old question as to how continents themselves are formed, something Wegener did not address. For him the continents were just assumed. By having two plates come together, the melted material produced tends to be more continental in character than the ocean crust rocks. If enough of this were produced, a small micro-continent could occur. Take these micro-continents and push them together, and you can build a large one. Simple, and, at first, this idea seemed to work well. Continents grew in size whether by the production of continental material at a subduction zone or by shoving the micro-continents together as a result of plate motion. However, in the last few years, as more detailed mapping has occurred, especially along the west coast of North America, we are finding that continental growth is not quite so easy to explain. From California to Alaska, geologists are finding the small micro-continents as expected, but with far more lateral movement than was expected. As many as 46 different slivers of continental material are found in western North America alone, and, based on their fossil content, some of these must have been deposited 2000-3000 kilometers south of where they are now. And to complicate the picture even more, they are finding slivers of the same micro-continent called Wrangellia scattered from Oregon to Alaska (Howell 1985). These pieces of continental material are well named "Suspect Terranes" (See Ben-Avraham 1981, Kerr 1983, and Howell 1985).

Finding these Suspect Terranes, for the moment, has not caused us to abandon the plate tectonics theory and its simplicity, because plate movement is still the best mechanism for transporting the micro-continents the distances required. What it has done is produce a good test of the concept, and these terranes have caused us to re-examine the simple continental accretion ideas that were an early outgrowth of plate tectonics. Only time and more data will tell us whether plate tectonics will go the way of Werner's Universal Ocean idea. Will the simplicity of plate tectonics augur well for its continuance as the reigning paradigm of geology? Will we, because the theory is so elegant and simple, allow it to linger beyond its useful life? Only time will tell.

University of Pittsburgh at Johnstown

REFERENCES

Bailey, E. B., 1967, *James Hutton: Founder of Modern Geology* (Elsevier, Amsterdam).

Ben-Avraham, Zvi, 1985, "The movement of continents," *American Scientist*, vol. 69, pp. 291-99.

Brice, William R., 1982, "Bishop Ussher, Lightfoot, and Creation," *Journal of Geologic Education*, vol. 30, pp. 18-24.

Clarke, John W. and Hughes, Thomas M., 1890, *The Life and Letters of the Reverend Adam Sedgwick*, Volumes I & II (Cambridge University Press, Cambridge, England).

Chamberlin, Thomas C., 1890, "The Method of Multiple Working Hypotheses," *Science* (Old Series), vol. 15, pp. 92-96 [Reprinted: 1965, *Science*, vol. 148, pp. 754-59].

Collet, Leon W., 1926, "The Alps and Wegener's Theory," *Geography Journal*, vol. 67, pp. 301-12.

Cox, A., Doell, R. R., and Dalrymple, G. B., 1963, "Geomagnetic Polarity Epochs and Pleistocene Geochronomery," *Nature*, vol. 198, pp. 1049-1051.

Cuvier, Georges, 1819, "Memoir of Werner": Reprinted in 1860 in *The Naturalists Library*, ed. by William Jardine, Vol. 29, (Publisher, Place), pp. 19-40.

Dietz, R. S., 1961, "Continent and Ocean Basin Evolution by Spreading of the Sea Floor," *Nature*, vol. 190, pp. 854-57.

Du Toit, A. L., 1937, *Our Wandering Continents* (Hafner Publishing Co., New York, NY).

Fenton, Carrol L. and Fenton, Mildred A., 1945, *Giants of Geology: the Story of the Great Geologists* (Doubleday & Co., Inc., Garden City, NY).

Feuer, L. S., 1957, "The Principle of Simplicity," *Philosophy of Science*, vol. 24, pp. 109-122.

Green, Mott T., 1982, *Geology in the Nineteenth Century* (Cornell University Press, Ithaca, NY).

Geikie, Archibald, 1905, *The Founders of Geology*, 2nd ed., (Reprint 1962) (Dover Publications, Inc., New York, NY).

Hess, Harry H., 1962, "History of Ocean Basins," in A.E.J. Engle *et al.* (eds.), *Petrologic Studies—A Volume in Honor of A. F. Buddington* (Geological Society of America, Boulder, CO), pp. 599-620.

Howell, David G., 1985, "Terranes," *Scientific American*, November 1985, pp. 116-125.

Hutton, James, 1788, "On the Theory of the Earth; or An Investigation of the Laws Observable in the Composition, Dissolution and Restoration of the Globe," *Transactions of the Royal Society of Edinburgh*, vol. 1, pp. 209-304.

Isacks, B. L., Oliver, J., and Sykes, L. R., 1968, "Seismology and the New Global Tectonics," *Journal of Geophysical Research*, vol. 73, pp. 5855-5899.

Kerr, Richard A., 1983, "Suspect Terranes and Continental Growth," *Science*, vol. 222, pp. 36-38.

Kemeny, J. G., 1953, "The Use of Simplicity in Induction," *The Philosophical Review*, vol. 62, pp. 391-408.

Laird, John, 1919, "The Law of Parsimony," *The Monist*, vol. 29, pp. 321-44.

Mayr, Ernest (ed.), 1952, "The Problem of Land Connections Across the South Atlantic Ocean with Special Reference to the Mesozoic," *American Museum of Natural History Bulletin*, vol. 99, pp. 79-258.

Morgan, W. J., 1971, "Convection Plumes in the Lower Mantle," *Nature* vol. 230, pp. 40-43.

Ospovat, Alexander M., 1969, "Reflection on A. G. Werner's Kurze Klassifikation," in Schneer, Cecil J. (ed.), *Toward a History of Geology* (The M.I.T. Press, Cambridge, MA), pp. 242-56.

----------, 1971, *Abraham Gottlob Werner, Short Classification and Description of the Various Rocks* (Hafner Press, New York, NY).

Pickering, W. H., 1924, "The Separation of the Continents by Fission," *Geological Magazine*, vol. 61, pp. 31-35.

Porter, Roy, 1977, *The Making of Geology: Earth Science in Britain 1660-1815* (Cambridge University Press, New York, NY).

Press, Frank, 1969, "The Suboceanic Mantle," *Science, vol. 165, pp. 174-76.*

Raitt, R. W., 1949, "Studies of Ocean-bottom Structure Off Southern California with Explosive Waves," *Geological Society of America Bulletin*, vol. 60, pp. 1915 (Abstract).

Runcorn, S. K., 1956, "Paleomagnetic Comparisons Between Europe and North America," *Geological Association of Canada Proceedings*, vol. 8, pp. 77-85.

Russell, Bertrand, 1929, *Our Knowledge of the External World* (W. W. Norton, New York, NY).

Snider, Antonio, 1858, *La Creation et Ses Mysteres Devoiles* (A. Franck & E. Dentu, Paris, France).

Taylor, Frank B., 1910, "Bearing of the Tertiary Mountain Belt on the Origin of the Earth's Plan," *Geological Society of America Bulletin*, vol. 21, pp. 179-226.

Tolstoy, Ivan and Ewing, W. Maurice, 1949, "North Atlantic Hydrography and the Mid-Atlantic Ridge," *Geological Society of America Bulletin*, vol. 60, pp. 1527-1540.

Vine, F. J. and Matthews, D. H., 1963, "Magnetic Anomalies Over Ocean Ridges," *Nature*, vol. 199, pp. 452-53.

Wadati, K., 1935, "On the Activity of Deep-focus Earthquakes in the Japan Islands and Neighbourhoods," *Geophysical Magazine*, vol. 8, pp. 305-25.

Wegener, Alfred, 1912, "Die Entstehung der Kontinente," *Petermanns Geographische Mitteilungen*, vol. 58, pp. 185-95, 253-56, 305-09. (Translation of excerpts in: Shea, James H. (ed.), 1965, *Continental Drift* (Van Nostrand Reinhold Co., New York, NY), pp. 147-64.

----------, 1924, *The Origin of Continents and Oceans* (translated from the 3rd German edition by J.G.A. Skerl) (E. P. Dutton, New York, NY).

Werner, Abraham G., 1774, *Von den Ausserlichen Kennzeichen der Fossiliern*; English translation 1962 by A. V. Carozzi, *On the External Character of Minerals* (University of Illinois, Urbana, IL).

Zittel, K. A. von, 1901, *History of Geology and Paleontology to the End of the Nineteenth Century* (W. Scott, London).

CONTRIBUTORS

WILLIAM R. BRICE
Department of Geology
University of Pittsburgh
at Johnstown

PAUL M. CHURCHLAND
Department of Philosophy
University of California
at San Diego

JANE DURAN
Department of Philosophy
University of California
at Santa Barbara

ULRICH MAJER
Philosophical Seminar
University of Göttingen

JOSEPH C. PITT
Department of Philosophy
Virginia Polytechnic Institute
and State University

NICHOLAS RESCHER
Department of Philosophy
University of Pittsburgh

DAVID B. RESNIK
Department of Philosophy
University of North Carolina
at Chapel Hill

KRISTIN SHRADER-FRECHETTE
Department of Philosophy
University of South Florida

MATTI SINTONEN
The Academy of Finland
Helsinki, Finland

NAME INDEX

NAME INDEX